T0332481

# INTRODUCTION TO ANALOG VLSI DESIGN AUTOMATION

## THE KLUWER INTERNATIONAL SERIES IN ENGINEERING AND COMPUTER SCIENCE

## VLSI, COMPUTER ARCHITECTURE AND DIGITAL SIGNAL PROCESSING

*Consulting Editor*
Jonathan Allen

**Other books in the series:**

*Logic Minimization Algorithms for VLSI Synthesis.* R.K. Brayton, G.D. Hachtel, C.T. McMullen, and Alberto Sanngiovanni-Vincentelli. ISBN 0-89838-164-9.
*Adaptive Filters: Structures, Algorithms, and Applications.* M.L. Honig and D.G. Messerschmitt. ISBN 0-89838-163-0.
*Introduction to VLSI Silicon Devices: Physics, Technology and Characterization.* B. El-Kareh and R.J. Bombard. ISBN 0-89838-210-6.
*Latchup in CMOS Technology: The Problem and Its Cure.* R.R. Troutman. ISBN 0-89838-215-7.
*Digital CMOS Circuit Design.* M. Annaratone. ISBN 0-89838-224-6.
*The Bounding Approach to VLSI Circuit Simulation.* C.A. Zukowski. ISBN 0-89838-176-2.
*Multi-Level Simulation for VLSI Design.* D.D. Hill and D.R. Coelho. ISBN 0-89838-184-3.
*Relaxation Techniques for the Simulation of VLSI Circuits.* J. White and A. Sangiovanni-Vincentelli. ISBN 0-89838-186-X.
*VLSI CAD Tools and Applications.* W. Fichtner and M. Morf, Editors. ISBN 0-89838-193-2.
*A VLSI Architecture for Concurrent Data Structures.* W.J. Dally. ISBN 0-89838-235-1.
*Yield Simulation for Integrated Circuits.* D.M.H. Walker. ISBN 0-89838-244-0.
*VLSI Specification, Verification and Synthesis.* G. Birtwistle and P.A. Subrahmanyam. ISBN 0-89838-246-7.
*Fundamentals of Computer-Aided Circuit Simulation.* W.J. McCalla. ISBN 0-89838-248-3.
*Serial Data Computation.* S.G. Smith, P.B. Denyer. ISBN 0-89838-253-X.
*Phonologic Parsing in Speech Recognition.* K.W. Church. ISBN 0-89838-250-5.
*Simulated Annealing for VLSI Design.* D.F. Wong, H.W. Leong, C.L. Liu. ISBN 0-89838-256-4.
*Polycrystalline Silicon for Integrated Circuit Applications.* T. Kamins. ISBN 0-89838-259-9.
*FET Modeling for Circuit Simulation.* D. Divekar. ISBN 0-89838-264-5.
*VLSI Placement and Global Routing Using Simulated Annealing.* C. Sechen. ISBN 0-89838-281-5.
*Adaptive Filters and Equalizers.* B. Mulgrew, C.F.N. Cowan. ISBN 0-89838-285-8.
*Computer-Aided Design and VLSI Device Development, Second Edition.* K.M. Cham, S-Y. Oh, J.L. Moll, K. Lee, P. Vande Voorde, D. Chin. ISBN: 0-89838-277-7.
*Automatic Speech Recognition.* K-F. Lee. ISBN 0-89838-296-3.
*Speech Time-Frequency Representations.* M.D. Riley. ISBN 0-89838-298-X.
*A Systolic Array Optimizing Compiler.* M.S. Lam. ISBN: 0-89838-300-5.
*Algorithms and Techniques for VLSI Layout Synthesis.* D. Hill, D. Shugard, J. Fishburn, K. Keutzer. ISBN: 0-89838-301-3.
*Switch-Level Timing Simulation of MOS VLSI Circuits.* V.B. Rao, D.V. Overhauser, T.N. Trick, I.N. Hajj. ISBN 0-89838-302-1.
*VLSI for Artificial Intelligence.* J.G. Delgado-Frias, W.R. Moore (Editors). ISBN 0-7923-9000-8.
*Wafer Level Integrated Systems: Implementation Issues.* S.K. Tewksbury. ISBN 0-7923-9006-7.
*The Annealing Algorithm.* R.H.J.M. Otten & L.P.P.P. van Ginneken. ISBN 0-7923-9022-9.
*VHDL: Hardware Description and Design.* R. Lipsett, C. Schaefer and C. Ussery. ISBN 0-7923-9030-X.
*The VHDL Handbook.* D. Coelho. ISBN 0-7923-9031-8.
*Unified Methods for VLSI Simulation and Test Generation.* K.T. Cheng and V.D. Agrawal. ISBN 0-7923-9025-3.
*ASIC System Design with VHDL: A Paradigm.* S.S. Leung and M.A. Shanblatt. ISBN 0-7923-9032-6.
*BiCMOS Technology and Applications.* A.R. Alvarez (Editor). ISBN 0-7923-9033-4.
*Nonlinear Digital Filters: Principles and Applications.* I. Pitas and A.N. Venetsanopoulos. ISBN 0-7923-9049-0.
*Algorithmic and Register-Transfer Level Synthesis: The System Architect's Workbench.* D.E. Thomas, E.D. Lagnese, R.A. Walker, J.A. Nestor, J.V. Rajan, R.L. Blackburn. ISBN 0-7923-9053-9.
*VLSI Design for Manufacturing: Yield Enhancement.* S.W. Director, W. Maly, A.J. Strojwas. ISBN 0-7923-9053-7.
*Testing and Reliable Design of CMOS Circuits.* N.K. Jha, S. Kundu. ISBN 0-7923-9056-3.
*Hierarchical Modeling for VLSI Circuit Testing.* D. Bhattacharya, J.P. Hayes. ISBN 0-7923-9058-X.

# INTRODUCTION TO ANALOG VLSI DESIGN AUTOMATION

edited by

**Mohammed Ismail**
Ohio State University

and

**José Franca**
Instituto Superior Técnico

**KLUWER ACADEMIC PUBLISHERS**
**Boston/Dordrecht/London**

**Distributors for North America:**
Kluwer Academic Publishers
101 Philip Drive
Assinippi Park
Norwell, Massachusetts 02061 USA

**Distributors for all other countries:**
Kluwer Academic Publishers Group
Distribution Centre
Post Office Box 322
3300 AH Dordrecht, THE NETHERLANDS

**Library of Congress Cataloging-in-Publication Data**

Ismail, Mohammed.
Introduction to analog VLSI design automation / by Mohammed Ismail and José Franca.
p. cm. — (The Kluwer international series in engineering and computer science. VLSI, computer architecture, and digital signal processing)
Includes bibliographical references.
ISBN 0-7923-9071-7
1. Integrated circuits—Very large scale integration—Design and construction. 2. Computer-aided design. 3. Neural computers.
I. Franca, José. II. Title. III. Series.
TK7874.I597 1990
621.39′5—dc20 90-4045
CIP



Printed in the United States of America

# Contents

**Preface** **vii**

**Chapter 1**
Analog Integrated Circuit Design Conceptualization 1
*Robert J. Bowman*

**Chapter 2**
Expert System for Parameterized Analog Cells 29
*Helena Pohjonen*

**Chapter 3**
Automated Design of Operational Amplifiers: A Case Study 45
*L. Richard Carley*

**Chapter 4**
Automation of Analog Design Procedures 79
*Georges Gielen and William Sansen*

**Chapter 5**
Analog Design Tools for High Frequency Applications 113
*Sorin Huss*

**Chapter 6**
Silicon Compiler Technology for SC Filters 151
*W. Martin Snelgrove*

**Chapter 7**
CAD-Compatible Analog System Design: A New Design Concept 163
*Mohammed Ismail*

**Index** **183**

# Preface

Very large scale integration (VLSI) technologies are now maturing with a current emphasis toward submicron structures and sophisticated applications combining digital as well as analog circuits on a single chip. Abundant examples are found on today's advanced systems for telecommunications, robotics, automotive electronics, image processing, intelligent sensors, etc.. Exciting new applications are being unveiled in the field of neural computing where the massive use of analog/digital VLSI technologies will have a significant impact.

To match such a fast technological trend towards single chip analog/digital VLSI systems, researchers worldwide have long realized the vital need of producing advanced computer aided tools for designing both digital and analog circuits and systems for silicon integration. Architecture and circuit compilation, device sizing and the layout generation are but a few familiar tasks on the world of digital integrated circuit design which can be efficiently accomplished by matured computer aided tools. In contrast, the art of tools for designing and producing analog or even analog/digital integrated circuits is quite primitive and still lacking the industrial penetration and acceptance already achieved by digital counterparts.

In fact, analog design is commonly perceived to be one of the most knowledge-intensive design tasks and analog circuits are still designed, largely by hand, by expert intimately familiar with nuances of the target application and integrated circuit fabrication process. The techniques needed to build good analog circuits seem to exist solely as expertise invested in individual designers.

In recent years, however, the scenario of computer aided tools for designing analog circuits and systems have shown signs of dramatic changes. Analog VLSI [1-3] has been recognized as a major technology for future information processing. Motivated by the reaches of methodologies and techniques of digital VLSI design automation, wealthy activity on analog VLSI design automation is underway. Albeit formidable obstacles to be successfully overcome, preliminary results are being unveiled which reiterate the correctness of the defined strategies and insuflate hope and encouragement for the work ahead.

This book is the first authoritive text dealing with the subject of analog VLSI design automation. It provides an excellent introduction to basic design automation ideas and their applications to development of knowledge-based systems and computer-aided tools for analog VLSI circuits and systems design.

The book is addressed to technically sophisticated readers and is recommended to all who are interested in VLSI design. It is a contributed volume presenting the combined efforts of leading researchers in the area of analog design automation. The book, which has grown out of lectures given at the 1988 IEEE International Symposium on Circuits and Systems in a workshop on *"Expert System Tools for Analog Signal Processing Applications"*[1] is divided into two parts. The first part, encompassing Chapters 1 to 4, deals with basic concepts of analog VLSI design automation whilst the second part, corresponding to Chapters 5 to 7, is divided to applications on such important areas in the microelectronics industry as high frequency bipolar technology and low frequency MOS technology.

Chapter 1, *Analog Integrated Circuit Design Conceptualization*[2] provides a formal description of the engineering design paradigm as well as a thorough understanding of the specific domain of analog circuit design. These are needed before machine-based emulation is possible, towards automating the human process of analog integrated circuit design, the steps of how does an engineer go from a system specification to a block diagram and then to an actual structure, are discussed and a model for the engineering design paradigm is presented. These early stages in the design are referred to as design conceptualization. The model is universal but the representation is specific to the domain of analog circuit design. Finally, a knowledge-based design strategy is introduced which mimics the human style for rapid analog circuit prototyping on an engineering workstation.

Chapter 2, *Expert System for Parameterized Analog Cells*[3], discusses methods for analog design and introduces the general components and

[1]The workshop was held in June 1988 at the Helsinki University of Technology, Helsinki, Finland and was sponsored by the Analog Signal Processing Technical Committee of the IEEE Circuits and Systems Society.

[2]Author: Dr. Robert J. Bowman, University of Rochester, U.S.A.

[3]Author: Dr. Helena Pohjonen, Technical Research Center of Finland.

structure of knowledge-based analog and mixed analog/digital expert systems. Hierarchy, abstraction and design styles of both analog and digital integrated circuits are addressed. It also addresses the need for, and provides means of, handling exact design and experimental information, interfacing with other tools like optimization and simulation tools and characterizing device and process parameters. Implementation aspects of artificial intelligence languages and expert shell systems are surveyed and possible limitations pointed out. The chapter concludes with descriptions of recently introduced analog expert systems and CAD tools such as PALLADIO, BLADES, AIDE2 and PROSAIC.

Chapter3, *Automated Design of Operational Amplifiers: A Case Study*[4], discusses design automation of one of the most important analog circuit building blocks, the operational amplifier (Op-Amp). OASYS: A prototype Op-Amp synthesis knowledge-based system is introduced. An organization for the system is first described. Analog circuit topologies are then represented as a hierarchy of functional blocks. A planning mechanism is introduced to translate performance specifications between levels in the circuit hierarchy. The system then provides schematic and layout of sized transistors for simple CMOS Op-Amps from performance specifications and process parameters. Circuit topologies as well as design equations are represented as statistically stored templates for use during the translation step. The material of this chapter will help reinforce design automation concepts discussed in the first two chapters.

Chapter 4, *Automation of Analog Design Procedures*[5], discusses an interesting view on the automation of the analog design procedures and stresses the important role of symbolic analysis tools. In a symbolic analysis program, the analytic expression of the performance is given in terms of device and/or process parameters. The role of each parameter in the expression is clearly recognized. As a result the designer learns about the performance of that circuit in a much more efficient way than running a circuit simulator such as SPICE. Moreover, the expressions can be used in a rule-based expert system in order to synthesize a new design. Symbolic analysis becomes even more important in the evaluation of unconventional characteristics such as power supply and common

[4]Author: Dr. Rick Carley Carnegie-Mellon University, U.S.A.
[5]Author: Dr. William Sansen, K. Universitet Leuven, Belgium.

mode rejections. The chapter introduces a new symbolic analysis tool suitable for both continuous- and discrete-time analog integrated circuits and discusses its applications. The analysis tool is then used in the creation of optimum-performance cell-based analog MOS VLSI library.

Chapter 5, *Analog Design Tools for High Frequency Applications*[6], focuses on automated design tools for high-speed linear and nonlinear bipolar analog integrated circuits in a frequency range up to 3 GHz. First, a characterization tool for high-speed bipolar devices is described followed by a discussion on high-performance high-speed analog cells. An optimization-base design tool suited to deal with complex circuits is presented followed by a description of a fully-hierarchical cell-based design package which supports extraction of parasitic resistive and capacitive components, thus allowing assessment of the implemented circuits. The tools described here are used at the Design Center for Integrated Circuits, AEG, West Germany. The chapter concludes with an overview of a rule-based design assistant which is actually being developed at AEG under an ESPRIT project. The design assistant is based on the concept of skeleton cells and takes a top-down design approach. Several design examples are included to demonstrate the main features of the tools.

Chapter 6, *Silicon Compiler Technology for SC Filters*[7], is devoted to the automatic synthesis of one of the most important classes of analog integrated circuits for MOS technology namely, MOS switched-capacitor (SC) filters. The chapter describes a prototype automated SC filter design and layout system called SICOMP. The system makes use of existing tools developed at the University of Toronto, Canada such as PREWARP: a program for prewarping attenuation specification, FILTOR2: a filter design package, AUTOSC: A SC circuit synthesis program, AUTOBLOCK: a block layout program and ALTOR: a standard cell automatic layout program. A menu driven manager program is responsible for calling the various programs in the system. A SC filter chip may be designed interactively or automatically using SICOMP. Design examples will be included with experimental results of a filter chip designed by SICOMP.

---

[6]Author: Dr. Sorin Huss, AEG Aktiengesellschaft, West Germany.
[7]Author: Dr. W. Martin Snelgrove, University of Toronto, Canada.

Chapter 7, *CAD-Compatible Analog System Design: A New Design Concept*[8] , provides an entirely new concept of automated analog circuit and system design. It advocates analog system design for CAD in a way that will bridge the gap between circuit design and CAD-oriented work in the analog field and bring analog VLSI system design a step closer towards automation. The new concept unveils strong and interesting connections between programmability, computer-aided design and physical implementation of analog VLSI circuits. This leads to system designs that rely on the interconnection of *simple* fixed analog cells instead of *parameterized* analog cells. The end result is a CAD-compatible area-efficient design. Examples of continuous-time analog MOS integrated circuits and systems are provided to demonstrate the new concept.

The authors would like to thank Robert Holland of Kluwer Academic Publishers for his assistance. M. Ismail is grateful to his graduate students, Nabil I. Khachab, Satoshi Sakurai and Seyed R. Zarabadi for their invaluable help in editing and preparing the final version of the book.

Mohammed Ismail.
Columbus, Ohio U.S.A.

Jose′ E. Franca.
Lisbon, Portugal

## References

[1] C. Mead, *Analog VLSI and Neural Systems*, Addison-Wesley, Reading, MA, 1989.

[2] C. Mead and M. Ismail, *Analog VLSI Implementation of Neural Systems*, Kluwer Academic Publishers, Boston, MA, 1989.

[3] M. R. Haskard and I. C. May, *Analog VLSI Design: NMOS and CMOS*, Prentice-Hall, NewYork, N.Y., 1988.

---

[8]Author: Dr. Mohammed Ismail, The Ohio State University, U.S.A.

# INTRODUCTION TO ANALOG VLSI DESIGN AUTOMATION

# CHAPTER 1

# ANALOG INTEGRATED CIRCUIT DESIGN CONCEPTUALIZATION

Robert Bowman
Department of Electrical Engineering
University of Rochester
Rochester, NY 14627
U.S.A

## 1 INTRODUCTION

The electrical signals of the natural world behave in an essentially continuous manner. In contrast, the world of electronic design, which seeks to process these continuous or analog signals, is predominantly discrete or digital. Motivated and refined largely by the evolution of the digital computer, digital design affords the electronic design community a powerful structured and portable design methodology. Substantial improvements in the productivity of the digital designer have been realized over the past decade through the use of computer–based design automation tools which leverage the design methodology. The effectiveness of digital design and test methods and the comparative lack of same in the analog domain have made digital architectures the dominant choice of system designers.As the limits in signal power, dynamic range, signal precision, and signal bandwidth are expanded the use of continuous signal or analog design methods becomes essential. The complicated architectures of proposed mixed analog/digital circuits and the inherent complexity of the analog design process simply overwhelm the capabilities of conventional design automation methodologies.

What is analog circuit design and what constitutes analog circuit design automation?

Analog circuit design is a human process that suffers from an identity crisis largely because of the convention of categorizing circuit design as either analog or digital. In fact, the recent preponderance of digital circuit design in the electronics community has resulted in a boolean definition for analog circuit design

$$ANALOG \equiv \overline{DIGITAL}$$

or anything which isn't digital design must be analog design. This simple expression highlights the difficulty of the analog design automation problem in terms of scope. The analog circuit domain extends from dc signals in power supplies to nanometer signals and beyond in optical sensors; from femptoamp currents in bioelectrodes to thousands of amps in control circuits; from radio astronomy receivers in liquid Helium at 4 K to CMOS pre-amplifiers attached to the end of an oil drilling bit at 500 F. All is covered by this convenient definition.

Therefore, the first steps in automating the process of "analog circuit design" are to identify the popular sub domains as performed in design practice and to determine if each sub domain design process can be formally described in sufficient detail for machine-based automation. The analog sub domains addressed in this text are those commonly associated with the design of application specific integrated circuits or ASICs. These sub domains are practiced by the majority of analog circuit designers and include bipolar and MOS monolithic integrated circuit design. We will refer to these sub domains as standard analog design practice.

Two logical goals of analog design automation for standard analog design practice are to:

1. significantly enhance the productivity of the experienced analog integrated circuit designer

2. expand the pool of designers that can participate in the process of analog integrated circuit design.

In this chapter, we examine the relevant issues in enhancing the productivity of analog integrated circuit designers through computer-based design automation methods. We will also formally describe analog integrated circuit design strategy as practiced by experienced human designers and concentrate specifically on design conceptualization or machine-based techniques for rapid analog circuit prototyping.

# 2 IMPORTANT ISSUES IN ANALOG DESIGN AUTOMATION

Analog design encompasses a myriad of distinct design disciplines that are loosely divided along common lines of thought such as bipolar versus CMOS or low frequency versus monolithic microwave. However, each discipline follows a generic human design process which is common to all analog integrated circuit design and perhaps most engineering design domains. It is this generic design process that must be understood and formally described before a universal automation paradigm for analog design can be realized. We will use low frequency CMOS analog integrated circuit design as a vehicle for describing some of these concepts.

## 2.1 Analog Design — Phenomenological Description

Let's begin by breaking down the overall design process into smaller tasks and determine where the design bottlenecks exist. Figure 1 summarizes the result of a survey of over 75 analog integrated circuit designers conducted by the author during the year 1988. The designers were employed at some 10 different companies like Analog Devices, Siemens Corporation and Tektronix and ranged in experience from 1–25 years.

The survey form ask each designer to estimate the percentage of total project design time (that time from specification to first hardware) devoted to each of 14 tasks. The tasks appear top to bottom in approximately the same sequence that they occur in a design process. The project size (number of active elements) and total project time varied greatly from designer to designer. However, within two groups of designers that we will designate as novice and experienced (greater than 3 years of design experience), the percentage time allocation for individual tasks was consistent throughout the survey .
Figure 1 organizes the 14 tasks into three design phases:

**I** Conceptualization (getting to a circuit topology)

**II** Optimization and Implementation

**III** Verification or Testing

Three observations about the analog design process can be made from the results in Figure 1. First, Phase III indicates that surprisingly little time is spent on testing and documentation. The results are a bit

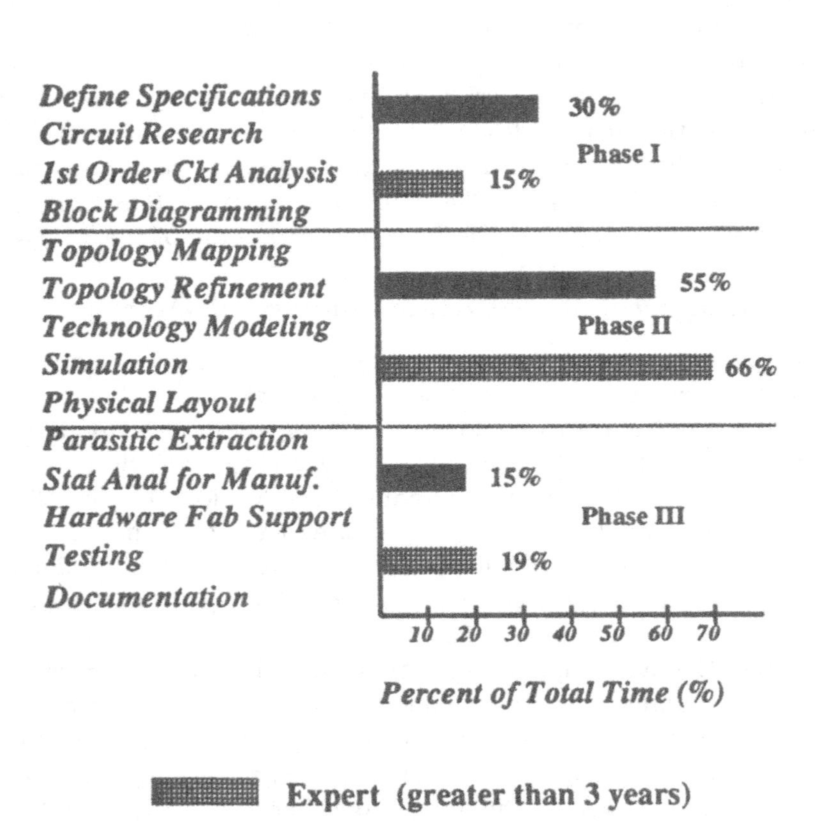


Figure 1: Analog Design Engineer Task Profile

misleading, however, as these tasks are commonly handled by someone other than the designer. In addition, many respondees stated a concern for the lack of formalism in analog testing procedures and the need to couple the test protocol development with the circuit design process. Analog test is certainly an area in need of attention by researchers, CAD tool developers, and test equipment manufacturers.

The dominant role of simulation and physical layout in the design cycle are evident as both novice and experienced designers devote considerable effort in Phase II. Many designers spend as much as fifty percent of the design time simulating circuit behavior. Physical layout is the major obstacle in effecting multiple design passes for optimization. If the goal of design automation is to enhance productivity, improvements in methods to perform these two analog design tasks is essential. Chapters Two, Four and Five discuss the development of some new tools to assist the designer in these areas.

Phase I emphasizes the difficulty novice designers experience in making the connection between electrical behavior and circuit topology or mapping from circuit function to circuit structure. Once a circuit topology is deemed suitable for a function the design procedure becomes much more methodical and uses well known analytical skills.

Figure 2 contrasts analysis and synthesis. Analysis applies well known circuit principles to estimate circuit electrical behavior and is the process taught to engineering students. Synthesis, the cornerstone of design, attempts to replicate electrical behavior by somehow arranging circuit primitives with known properties in a suitable structure. The synthesis paradigm involves analysis but is much more than analysis. Synthesis is an ensemble of answers waiting for the right question. Those designers with a rich design experience have filled the solution space with good approximations to most questions. However, the general synthesis paradigm is not well understood and is virtually impossible to teach using a classroom pedagogy. As a result, novice designers will always experience difficulty in realizing circuit topologies. Design automation for analog must be sensitive to differences in the various circuit classes within a sub domain as well as the sub domains themselves. Figure 3 portrays relative design complexity for nine different monolithic integrated circuit classes. Design complexity is a somewhat subjective term but attempts to indicate the relative difficulty encountered by most designers when faced with circuit problems in the classes shown. In most instances, a circuit class is considered complex if the synthesis process

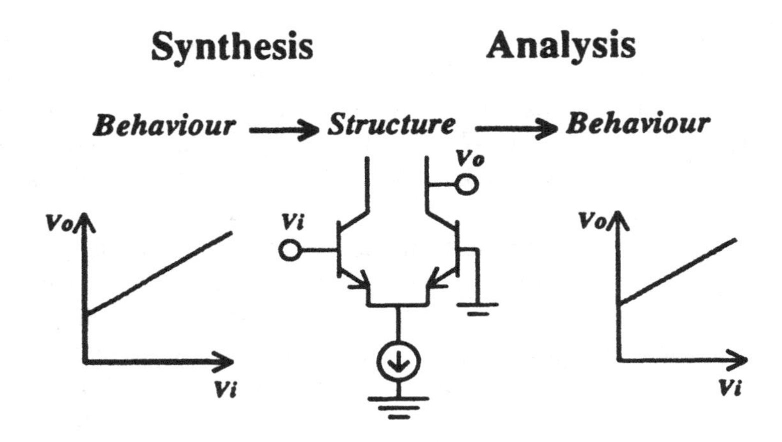


Figure 2: Circuit Analysis versus Circuit Synthesis

requires considerable non-linear analysis, precise technology modeling, or an unusual circuit topology.

# 3 CHARACTERIZING ANALOG INTEGRATED CIRCUIT DESIGN

Comparatively little attention has been devoted to automating the design methodologies for analog integrated circuit composition. Some of the early research into knowledge-based analog circuit design concentrated on representing design knowledge at the circuit element level and on qualitative reasoning about structure and function in discrete element circuits [1,2,3]. This approach mimics the reasoning process of a novice circuit engineer and may ultimately provide the most complete paradigm for knowledge-based design. But the inherent non-linearity of a single active element and the many considerations in establishing dc bias for common circuit topologies, engenders a design automation task of enormous scope and complexity. Recent work [4,5] has investigated the use of machine assisted selection of previously designed analog macrocells.

In this chapter we attempt to characterize analog integrated circuit design strategy as practiced by many experienced human designers, and propose a knowledge-based system configuration to automate this pro-

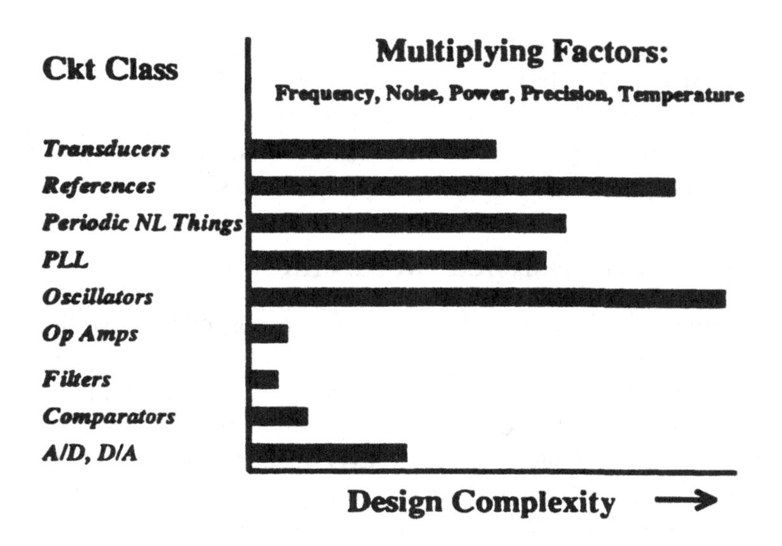


Figure 3: Analog Design Complexity and Circuit Classes

cess. A scaled model of this knowledge-based system has been developed to demonstrate machine automated analog integrated circuit design for a restricted class of circuits [6]. A data base structure which contains circuit performance, topology, and function attributes is described. Data base elements are represented as circuit frames in three levels of complexity to allow rapid searching of the data base and to parallel human design strategy.

We wish to emulate the human process by which analog integrated circuits are synthesized or analyzed. Assuming the design domain is both definable and programmable it is necessary to describe the design process in sufficient detail so that machine-automated reasoning may be applied. In this section we detail commonly used circuit design principles (tenets) and present the pattern of reasoning followed in a typical design session. In a later section we examine the significance of these principles when represented in a knowledge-based system.

# 4 DESIGN TENETS

Advanced analog integrated circuit design philosophy adheres to no single style. The truly creative results of advanced design appear to transcend the combined human talents of scientific method and technological awareness. Of course, circuit designers rely heavily on circuit analysis skills such as large and small signal analysis, non-linear device modeling, and integrated circuit process statistical analysis. In practice, advanced designers follow proven design principles or tenets.

## 4.1 Circuits are designed hierarchically

Figure 4 depicts the hierarchical nature of analog circuit configurations with a typical circuit example of each level indicated at the right. The person referred to as a "circuit designer" practices his profession on circuit projects. Circuit projects are typically specified at the module level and above in complexity. The early design strategy concentrates on clarifying and prioritizing specifications and partitioning the project into small functional blocks of complexity comparable to the stage level. At the stage level the analytical problems are tractable. Large and small signal circuit analysis is practiced predominantly at the stage level. The designer attempts to apply optimized stages from previous design sessions or alter those stages where necessary to synthesize the complete circuit.

Stage performance cannot always be satisfied by applying stage configurations from past experience and a new stage (for the designer) must be synthesized. In contrast, some projects may be mapped immediately into more complex circuit topologies than the stage level. For example, some active filter configurations are readily adapted to new applications with only minor alterations.

## 4.2 Designers accumulate a bag of circuit tricks

Throughout a designer's career, many functional circuit blocks are analyzed, categorized according to task, and stored away for possible future use. The experienced designer will accumulate a "bag-of-tricks" or specific circuit configurations with particularly appealing performance traits that satisfy a host of design problems with little modification. As a result, the process of circuit synthesis for experienced designers is greatly enhanced. In addition to synthesis, a large repertoire of well understood circuit blocks considerably simplifies the understanding of circuits designed by others as many of these blocks are employed by the general design community. Most of advanced circuit analysis is really pattern

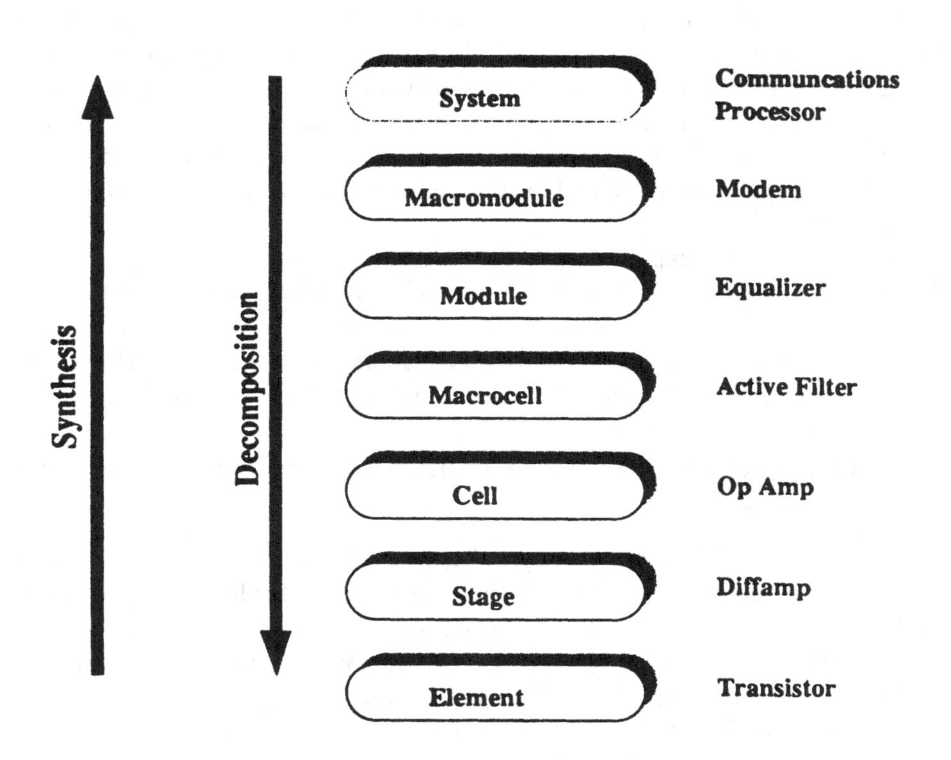


Figure 4: Analog Circuit Hierarchy

recognition and performance recall at the stage or cell level of circuit hierarchy.

An example of a commonly used optimized building block, the band gap voltage reference stage is shown inside the highlighted window of the temperature compensated voltage reference in Figure 5. The synthesis and optimization of this simple but effective circuit topology has transpired over years of work in several technologies. Yet once a designer determines this stage level element configuration as appropriate for a new task the design proceeds immediately from circuit synthesis to optimizing a known circuit configuration for a specific application. Combined with another common circuit topology , the op amp, the band gap cell provides a temperature compensated voltage reference. Note the simple hierarchy in Figure 2 with op amp A1 represented as a simple icon.

Once the band gap voltage reference circuit stage is thoroughly understood it is useful in many different circuit applications with minor variation. Conversely, once the circuit topology is recognized in a previously designed circuit as a reference, it is useful to visualize this stage as a generic model of a voltage source to simplify circuit analysis.

## 4.3 Designers use rules-of-thumb to simplify the design process

As with most design domains qualitative or commonsense reasoning comprises an essential part of circuit design philosophy. Years of integrated experience in analysis and synthesis with consideration to technology are concentrated into a set of highly efficient, easily applied heuristics. Each of the following demonstrates a rule or set of rules-of-thumb as applied to the circuit synthesis and analysis.

- the voltage drop between base and emitter junction of a silicon bipolar transistor, is about 0.6 volts.

- In a bipolar input differential stage the level of collector bias current is set to satisfy competing specifications of high slew rate and low input bias current. Both slew rate and input bias current are proportional to collector bias current. If neither parameter is a dominant specification each will be compromised by setting collector bias to optimize transistor beta and thus the first order frequency response of the complete stage.

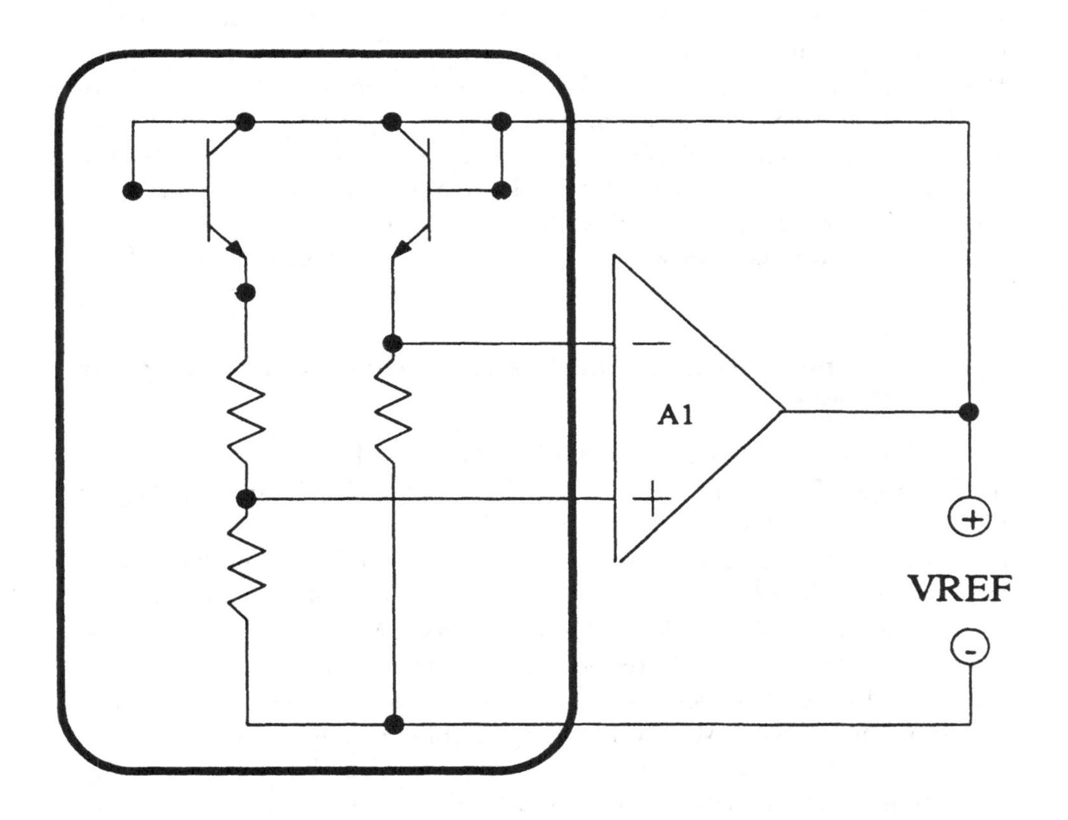


Figure 5: bBand Gap Stage within a Temperature Compensated Voltage Reference

- MOS integrated circuit capacitors will fall in the range of 0.01 to 50 pf for most designs due to area constraints and surrounding parasitics.

- Small signal gain in a CMOS differential amplifier decreases with increasing drain bias current.

Each of the above rules constitutes a circuit design heuristic, a circuit characteristic found reliable in most settings. The first rule is valid if the transistor is biased in the forward active region. However, there are instances when the base-emitter junction is reverse biased for use as a zener reference resulting in a 6 volt potential difference across the junction and with opposite polarity. Each rule is in someway linked to physical properties but the suitability of applying each in design judgment varies with the specific application and the experience level of the designer.

## 4.4 Designers optimize and retain tricks at the stage or cell level

Human designers operate comfortably with ensembles of active elements in the range of six to ten elements or less. Most circuit analysis begins after partitioning larger circuits to this level of complexity. In terms of a minimum element configuration, the relation between analog circuit function and circuit structure contributes to the preference by experienced designers to synthesize at the stage level. Many analog integrated circuit performance attributes are not realized until a complete bias "chain" is formed between the supply rails (denoted bold). As shown in Figure 6 the simple CMOS differential amplifier consists of 6 elements when one includes the current mirror used to bias the stage. Viewing the unique performance attributes of a differential amplifier such as common mode range or small signal differential gain requires the minimum topology of 6 elements.

## 4.5 New stage designs are rarely novel

Rarely does a new stage design result in a truly novel circuit configuration such as the four quadrant multiplier. "New" in this sense means new to the design team but not necessarily new to the analog design domain. Application and technology may impose new production rules for

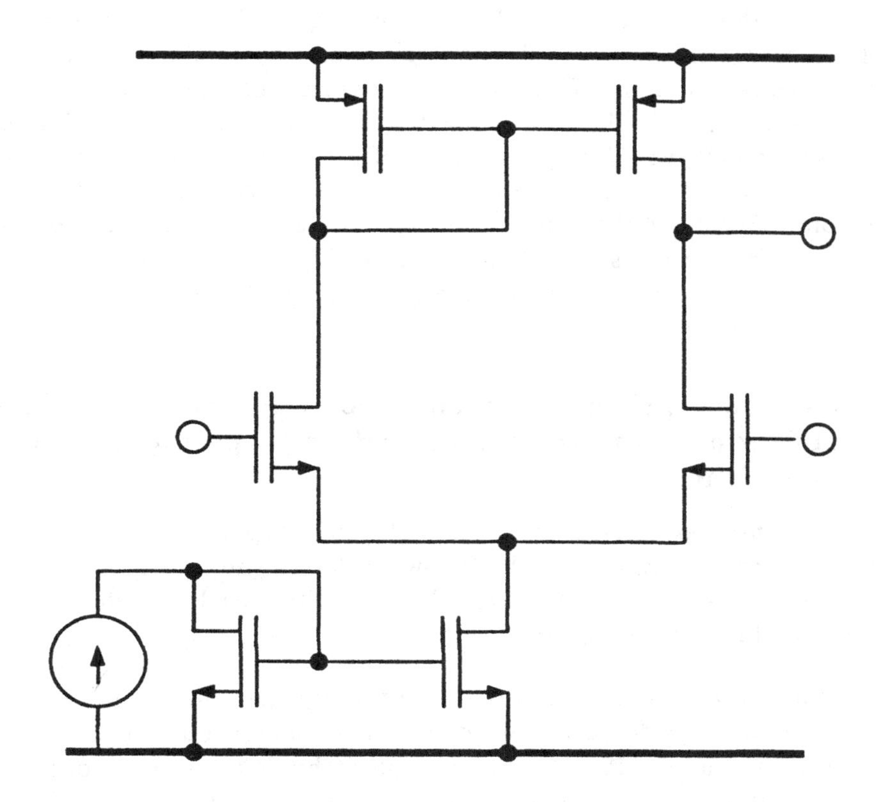

Figure 6: Simple CMOS Differential Amplifier

the design of a stage but in most instances the transfer characteristics, input/output definitions, and even circuit topology will closely resemble a previously designed stage. This highlights the importance of retaining and distributing stage level design knowledge.

## 4.6 Stage topologies are replicated in technology

Figure 7 shows a differential amplifier implemented in Bipolar technology.. Notice the identical active element topology with that of the CMOS amplifier shown in Figure 6. Technology analogies at the stage or cell level are quite common. The relation between structure and function becomes apparent. Design reasoning tends to track circuit topologies. Analysis of a current mirror, whether in Bipolar or MOS technologies, follows the same train of thought. The mathematics may be slightly more complicated for the bipolar case but the essence of the reasoning process still concerns itself with the relationship of reference current to output current and quality of the arrangement as a current source/sink.

## 4.7 Designers prioritize specifications in terms of achievability and importance according to previous design experience

The design engineer carefully assesses each project specification based on previous experience. Consider the following design judgments in achieving specifications for a single mirror-loaded differential amplifier stage (circuit topology of Figure 6):

- Realizing a small signal differential voltage gain minimum of 2000 would be considered standard practice using Bipolar technology but the same specification in a CMOS technology would be deemed very difficult due to much lower stage transconductance in MOS technology.

- A specification for a minimum common mode rejection ratio of 50 dB at 1 Khz would be assessed as very achievable in either technology based on typical gain and matching properties.

- In CMOS technology achieving both a small signal gain of 200 and a 1 milliamp drain bias current to support high slew rate would be

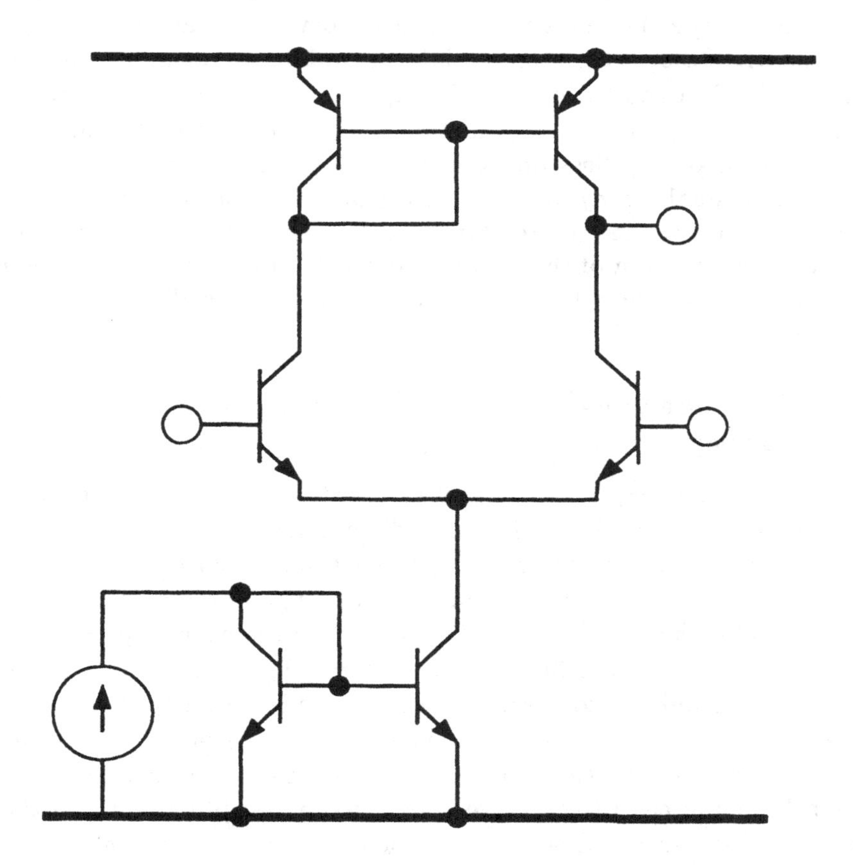

Figure 7: Simple Bipolar Differential Amplifier

ranked as very difficult whereas this would be standard practice in Bipolar technology.

- Maximum input bias current of 100 pA is a reasonable specification in CMOS but not in Bipolar.

Inevitably the designer is confronted with fundamental design tradeoffs: those circuit specifications whose improved values track in opposite directions with respect to design variables. Small signal gain is enhanced at the expense in bandwidth. Circuit impedance is reduced with increases in input bias current. In these instances the specifications compromised are those ranked as "less important" or if the competing specifications are ranked equally they are both compromised in a manner to optimize another highly ranked parameter. Design strategy is formulated on the designer's perception of the relative degree of difficulty in achieving each specification and the relative importance of each specification in the total project.

## 4.8 Few specifications strongly influence early design strategy

Early design strategy concentrates on partitioning the specification according to function and is dominated by a few high priority specifications. Experienced designers reduce the list of total project specifications to several design keys as a means of mapping function into structure. For example, the complete specification for an active filter might consist of as many as 30 different parameters. The one parameter of 160 dB/decade rolloff is recognized as a *design key* in that the order of the required transfer function to achieve this parameter leads to one of several logical circuit architectures. In many instances the connection between design keys and circuit structure is not so obvious but the use of the design keys steer the processes of decomposition and circuit synthesis along optimal paths. A low noise system specification would focus attention on the signal entry points in the circuit as system noise is dominated by front end performance.

Therefore, the designer makes the circuit project tractable by focusing on a few important specifications or designs keys. The design keys steer the decomposition and synthesis strategy until sufficient circuit detail is realized to permit consideration of the entire list of specifications. Many of the secondary specifications will be satisfied after the design approach meets the performance requirements of the design keys because of the interdependent nature of many circuit specifications.

# 5 AN ANALOG CIRCUIT DESIGN PARADIGM

In composing modern analog integrated circuits the advanced designer follows a distinct pattern of reasoning. Figure 8 depicts a reasoning process which emulates the human designer. The blocks represent key steps in a typical design session with design iteration shown as feedback arrows. Mainstream reasoning is supported by skills denoted as circles. These skills reflect the experience level of the designer in three areas: judgment in assessing specifications, knowledge of circuit topologies, and circuit analysis skills. The objective for any design session is the realization of a circuit configuration that satisfies the specifications as interpreted by the designer. The process of **Interpretation** translates specifications into quantitative forms: Declarative Exact (Input=differential), Declarative Inexact (Input Impedance > 100Kohms), Nominal (25 C), Nominal-Tolerance (5 Volts ±10%), Range (100 na < Input Bias Current < 200 na), Conditional (CMRR=50dB at 1 KHz), and combinations of these forms. This step also transforms the specifications into unambiguous terms meaningful to the designer. **Assessment** of the specifications assigns a priority to each for purposes of steering design strategy along optimal paths. For example, a 10 bit analog-to-digital converter is required to have a 50 MHz sampling rate. The 50 MHz sampling rate would be assessed a high priority and steer the strategy away from slow technologies (NMOS, CMOS) and slow methodologies such as dual slope integration and successive approximation. Performance data from previous design sessions (experience) dictate whether a specification is assigned as a *design* key (high priority objective).

The design keys that emerge from the prioritization dominate the early decision making in a circuit design session. In a CMOS opamp design with a high priority on large differential gain, differential stage bias current is reduced and slew rate compromised to achieve high gain. In addition, the design keys assist in the **Decomposition** process to arrive at a functional block diagram. In the CMOS opamp design, the high gain design key would invoke decomposition along lines of architectures with cascaded gain stages.

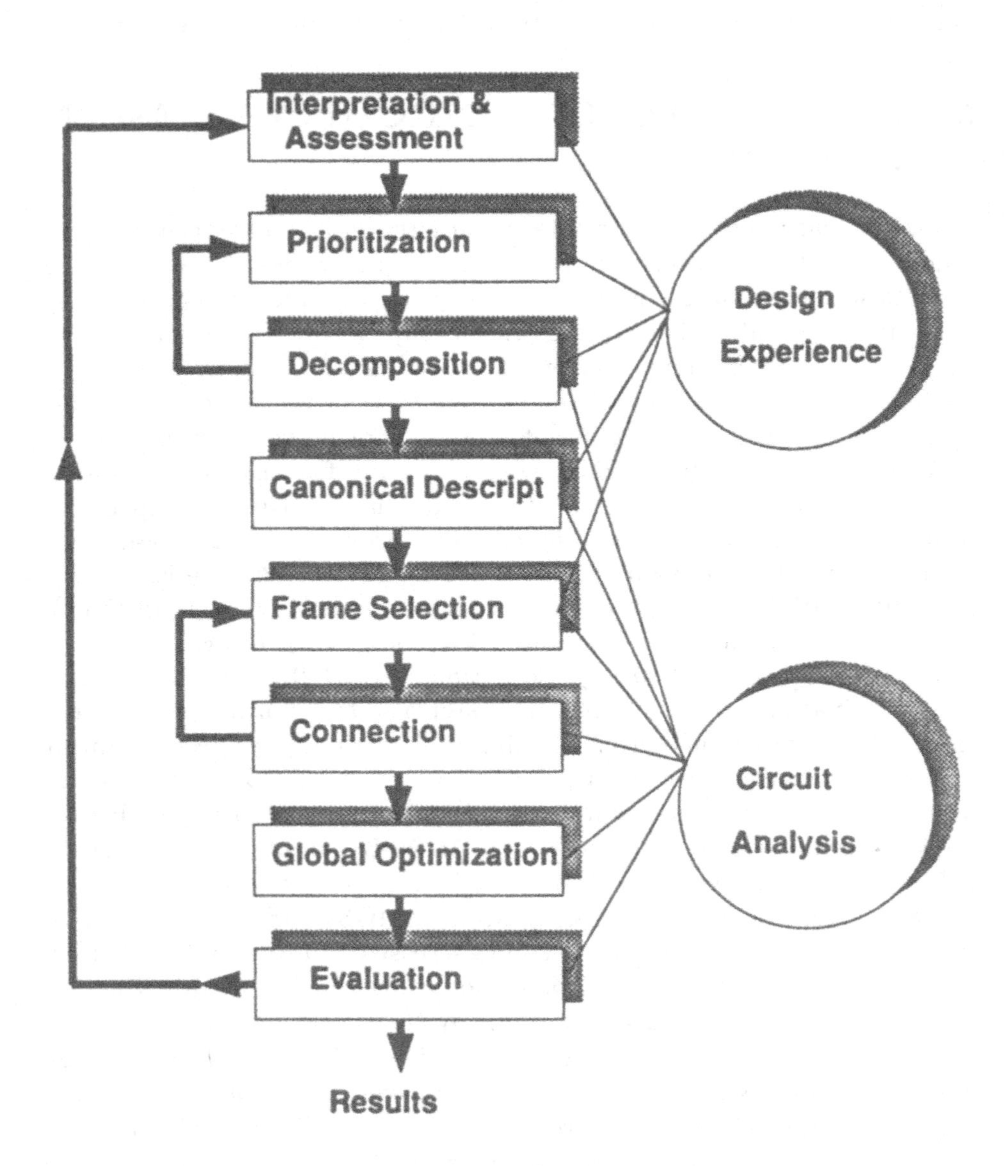


Figure 8: Circuit Design Paradigm

**Logical blocking** maps the project specification into performance blocks with loosely defined circuit topology. Blocking attempts to partition the project into circuit frames, or analog circuit topologies that usually perform as well in the context of a surrounding network as they are isolated. This tends to be a difficult process and often requires iteration with decomposition.

Candidate circuit frames are searched during **Frame Selection** for correspondence in port equivalency and nominal performance. Internal circuit frame topology is not considered until candidate frames match in port configuration, port type and circuit function. Circuit function is the transfer function associated with the principal signal path as designated by the design keys. For example, a design key of low noise would relate to a circuit function of noise figure between signal ports denoted input and output.

The process of selecting circuit frames proceeds until all candidate circuit frames at the port level and block function are found. The performance evaluation begins with the circuit frames being represented as specific topologies to determine nominal performance data. At this point circuit frames consist of circuit configurations but not specific transistor sizes or bias current. Each circuit frame is characterized by a range of performance data resulting from moderate scaling of the design variables (transistor size and bias current).

The heuristic process of realizing an acceptable circuit frame schematic from nominal performance data relies heavily on the design keys (tenet 8) adopted for that design session. Whether or not a circuit frame can satisfy a list of performance specifications is a rather complex function of how the specifications are prioritized and the conditions under which the circuit frame's performance was characterized. Design keys or high priority performance specifications stipulate how the frame is to be evaluated for nominal performance of that topology. For example, the differential amplifier in Figure 6 optimized for differential signal bandwidth will result in transistor sizing substantially different from one optimized for low noise.

The synthesis of a complex circuit involves the **Connection** of circuit frames in accordance with the blocking strategy. Interaction of circuit frames alters the nominal analog performance of individual frames. To gauge the effects of interaction and to adjust interport biasing, the connection process applies conventional circuit analysis methods such as first order hand analysis or the use of a circuit simulator and sensitivity analysis to construct a connection rule base. Interconnection of circuit frames alters the nominal analog performance of individual cells. To gauge the effects of interaction and to adjust interport biasing, the connection process applies conventional circuit analysis methods such as first order hand analysis or the use of a circuit simulator. In most instances interconnection of two circuit cells can only be accomplished after adjusting dc bias levels on connected signal ports for equivalency. The rules chosen for optimal adjustment of the dc bias should attempt to minimize degradation of desired signal performance in those macrocells.

Consider a rule set for implementing dc bias adjustment on ports A and A' for two CMOS frames (see Figure 6): the differential amplifier and a simple gain stage intended for linear circuit applications. The rule set assumes a design key of small signal ac gain.

1. **Ports are eligible for connection if their isolated dc bias values reside in the same supply corridor.**The power supply voltage range (VDD - VSS) is divided into two equal corridors (C+ and C-) where [VDD-VSS]/2 is the dividing line. In MOS technologies where complementary devices are available gate bias voltages for p type devices typically lie in C+ and n devices in C-. Input ports of cells are connected predominately to transistor gates. Stipulating that port A and port B bias potentials reside in the same corridor enhances the achievement of a dc bias match with minor variation of design variables and probably eliminates the need for level shifting.

2. **Design variables (DV) available to match port dc bias are transistor width (W), transistor length (L), and bias current (Ib).** Bias current is usually adjusted by W and L also but in some cases a separate resistive element is used for negative feedback.

3. **Design variables are altered within a range of DV/n $<$ DV $<$ nDV where n is a scale factor.** n is typically a value between 1 and 3 and whose upper bound is chosen to accommodate the IC layout scaling system used and the circuit frames sensitivity to behavior degradation.. Values greater than 3 tend to alter circuit frame analog performance too much and require an associated coarse grid layout generator which is inefficient in the use of silicon.

4. **Design variables (DV) are altered in a sequence of W, L, W and L, Ib.** This sequence alters first those design variables whose influence on important analog circuit attributes is the smallest. In this instance, a key performance parameter is small signal gain $Av$ described by the first order relation $Av = gmxro$ where $gm$ is the stage transconductance and $ro$ is the small signal impedance driven by the stage. Expressing $gm$ and $ro$ in terms of the design variables and determining the sensitivity of $Av$ to each design variable leads to the sequence given above. In cases where more than one design key is dominant a multivariate sensitivity analysis is appropriate.

5. **Driven circuit frame (II) is altered first according to (4), then the driving frame (I), then both in attempting to achieve a bias match on ports.** This sequence is based on the premise that the design is progressing in the direction of the signal flow and therefore presumes the next addition in the circuit design to be the driven stage. Altering the new stage (II) first will minimize degradation of previously achieved performance traits. Of course, this premise is not acceptable for all designs such as those involving feedback or those that include load impedance as a dominant design key.

**Global Optimization** attempts to optimize composite designs. In many instances tradeoffs at the stage level are possible and desirable after a circuit is synthesized to a higher level in the hierarchy. For example, a three stage MOS op amp is determined to easily satisfy a slew rate specification but is slightly below spec on overall gain. At this point it may be possible to compromise slew rate for additional gain to achieve the specification.

Ultimately the composite circuit must satisfy the total specification. The **Evaluation** of the all "known" circuit structures using the design strategy imposed by the specification may result in no suitable solutions. A reassessment (relaxing or change in priority) of the specifications permits one to invoke design iteration.

Successful circuits are represented as schematic diagrams and composite layouts. **Results** for each circuit provide a basis for selecting the best circuit of multiple choices according to some secondary strategy (power dissipation, chip area, etc.).

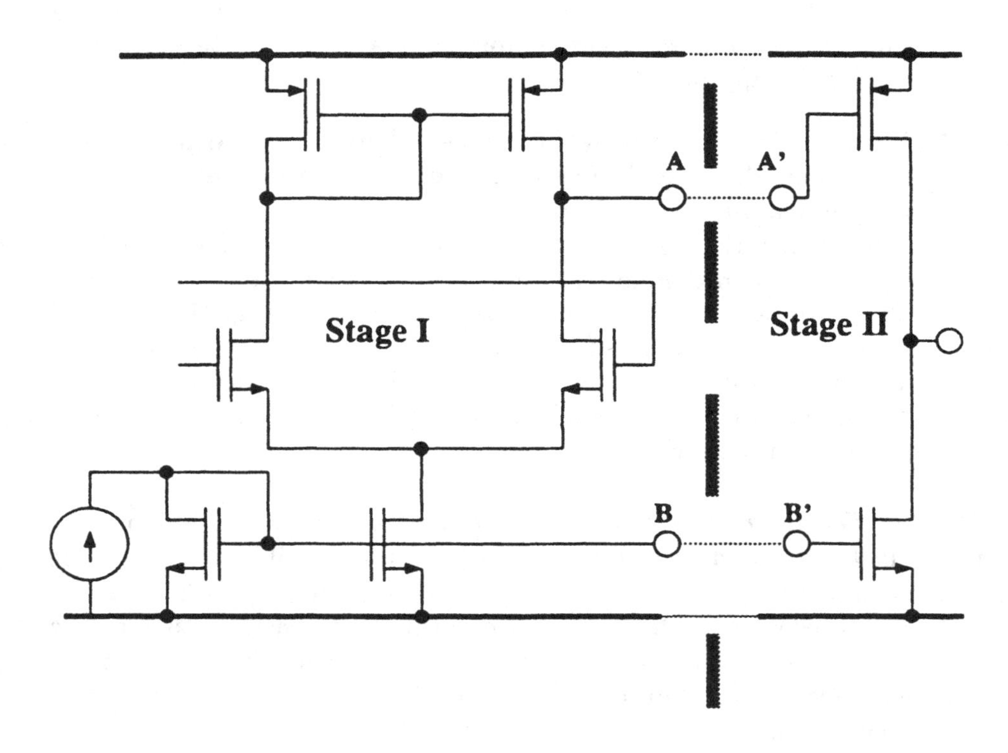


Figure 9: Connection of Two CMOS Analog Macrocells

# 6 ANALOG DESIGN DOMAIN REPRESENTATION

Is it feasible to represent analog circuit design knowledge for machine -based reasoning in a way that sufficiently emulates actual design practice using today's machine (computer) technology? Let us examine the design tenets and the pattern of reasoning of the human designer described in the previous section to deduce a suitable data base structure and design knowledge representation.

Paramount to the advanced design process is circuit abstraction at the stage or cell level in circuit element complexity. The analog circuit hierarchy for advanced designers as indicated by tenets 1, 2, and 4 includes both the stage level and the transistor level. A representation scheme for circuit configurations must support this notion. Certainly a stage-level data base representation is more suited than component-level to implementation in a knowledge-based system in terms of minimizing complexity. Stage-level representation speeds early synthesis as the knowledge structure need not concern itself with elementary but computationally intensive considerations such as arranging internal stage topology or overall stage-level large signal bias.

Advanced designers packages function and structure into higher level abstracts. The data base of circuit frames must be organized similarly. Initial rapid searching of the data base for candidate circuit frames should be performed on a very cursory level or the highest level of abstraction. As the design session progesses more detailed information on eligible circuit frames is sought by examining lower levels of abstraction. The data base is descriptive in nature much like an IC data catalog but it also represents design knowledge implicitly. The mere existence of a circuit stage topology in the data base implies that it is a useful arrangement in some application.

The importance of abstracting circuit frame information becomes evident when one considers the search complexity of a modest data base consisting of 80 distinct frames. Each frame is characterized for nominal signal and bias performance for 10 different design keys and 10 different bias levels. Assuming only 5 minutes of simulation time are required for each case leads to over *27 days of computer time* to determine the best frame.

These frequently used circuit configurations, referred to as circuit frames, can be thought of as self contained building blocks. In actual circuit design these blocks are starting points for synthesis or analysis. For example, the circuit topology in Figure 3 is immediately recognized as a differential amplifier. The differential amplifier is a universal building block for any analog designer and may be described in terms of its nominal performance characteristics under constraints of technology, bias current range, and operating conditions. Placing this circuit frame in a circuit network or in context may result in significant changes in the circuit frame's small and large signal performance and require corrections in dc bias to satisfy port connection. However, already knowing the circuit topology permits the circuit analytical methods to proceed rapidly.

From tenets 5 and 6 the importance of a circuit data base becomes evident. The quality of future designs depends heavily on the ability to recall and invoke known circuit structures. In addition, recall should be capable of crossing technology boundaries.

Design strategy is formulated on the assessment of specifications. The data base must provide data structures which accommodate specification data, priorization of this data, and design keys to simplify the early design strategy.

In summary, the data representation approach must support:

1. circuit hierarchy
2. circuit frame structure or topology
3. multi-level circuit frame abstraction
4. circuit frame function
5. circuit technology
6. design specifications and design keys

A logical view of an appropriate circuit frame data base structure is depicted in Figure 7. An individual macrocell appears in three levels of abstraction in the data base following the trend of general to specific from top to bottom. The top or profile level is the most abstract. Circuit topology is evident at the second level with the actual circuit schematic resulting on the third level. As the design session progresses more detailed information on eligible circuit frames is sought by examining lower levels but at greater computational expense.

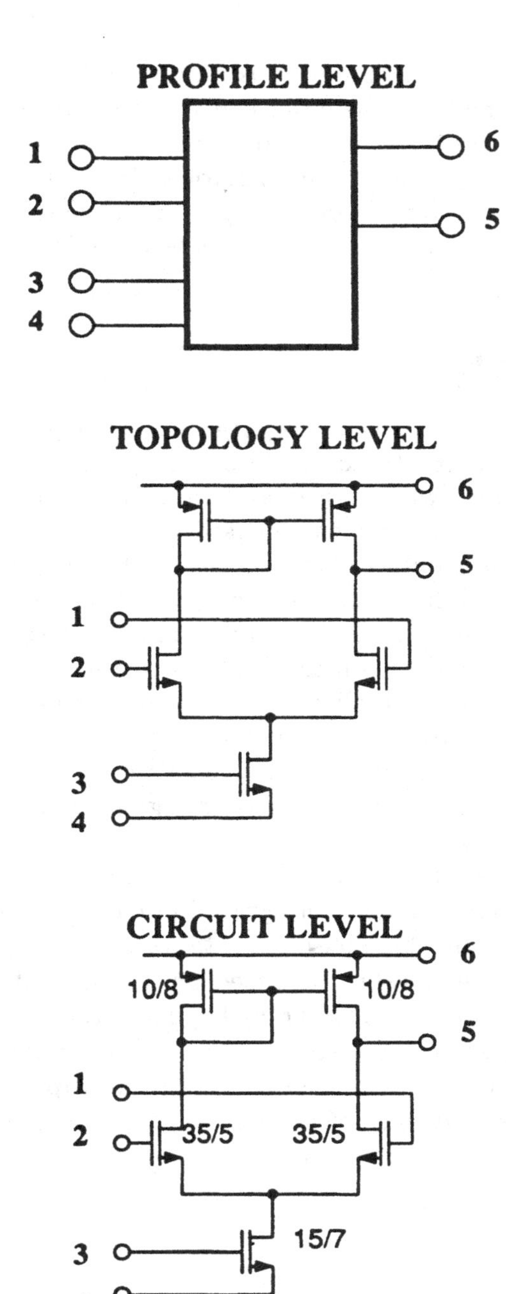


Figure 10: Multi-level Representation of Circuit Frames

The circuit topology in Figure 7 is immediately recognized as a differential amplifier and may be described in terms of its nominal performance characteristics under constraints of technology, bias range, and operating conditions. Placing this macrocell in a circuit network or in context may result in significant changes in the macrocell's small and large signal performance and require corrections in dc bias to satisfy port connections.

# 7 SUMMARY

Design automation for analog integrated circuit engineering is complicated by the lack of structured, portable design methods and the broad spectrum of design knowledge. Research into enhancing the productivity of the analog circuit designer has focused primarily on the sub domain of low frequency monolithic integrated circuits due to the significant demand for analog circuit functions on application specific ICs.

Determining a suitable circuit structure in the design of analog integrated circuits remains as a difficult task for systems level analog designers. Analog design conceptualization deals with the early phases of the circuit design process to the point where an appropriate circuit topology is realized. Design automation in this area attempts to develop machine-based techniques that assist the analog designer in rapidly determining an analog circuit topology.

In this chapter a phenomenological description of the human process of designing monolithic analog integrated circuits and a representation of this paradigm for machine-based assistance has been presented. A decomposition/synthesis strategy is based on a search of commonly used circuit frames. Circuit frame hierarchy begins at the stage level of circuit complexity for most design sessions, thus considerably simplifying the design task. Circuit frames are represented in multiple levels of abstraction to improve knowledge space searching and to parallel the design reasoning of experienced designers.

# References

[1] G. J. Sussman and G. L. Steele, "CONSTRAINTS—A Language for Expressing Almost Hierarchical Descriptions," Artificial Intelligence, vol. 14 pp. 1–39, 1980.

[2] J. DeKleer, "How Circuits Work," Artificial Intelligence, vol. 24, pp. 205–280, 1984.

[3] J. DeKleer and G. J. Sussman, "Propagation of Constraints Applied to Circuit Synthesis," Circuit Theory and Applications, vol. 8, pp. 127–144, 1980.

[4] M. G. Degrauwe and W. M. Sansen, "The Current Efficiency of MOS Transconductance Amplifiers," IEEE Journ. Solid State Ckts., vol. sc-19, pp.349–359, June , 1984.

[5] P. E. Allen and H. Navarez-Lozano, "Automated Design of MOS Op Amps," IEEE Symposium on Circuits and Systems, pp. 1286–1289, 1983.

[6] R. J. Bowman and D. J. Lane, "A Knowledge-based System for Analog Integrated Circuit Design," IEEE Int. Conf.on Comp Aided Design, vol. ICCAD-85, November, 1985.

[7] R. J. Bowman, "A Rule Based Approach for Adapting Analog Macrocells to Specified Circuit Contexts," Proceed. 30th Midwest Symposium Ckts and Systems, August , 1987.

[8] R. Harjani, R. A. Rutenbar, and L. Rick Carley, "A Prototype Framework for Knowledge-based Analog Circuit Synthesis", Proceed. 24th Design Auto. Conf., June 1987.

# CHAPTER 2

# EXPERT SYSTEM FOR PARAMETERIZED ANALOG CELLS

Helana Pohjonen
Technical Research Center of Finland
Semiconductor Laboratory
Olarinluoma 9, 02200 Espo
Finland

Designing analog functions means different design style, synthesizing and optimizing functions compared to the digital design. These differences have raised research topics of combining new artificial intelligence (AI) tools, synthesizing, optimization methods and the simulation iteratively.

This chapter discuss methods and introduces the general components and structure of knowledge based systems available for the analog and analog/digital design. It also address the need for, and provides means of, handling exact design and experimental information, interfacing with other tools like optimization and simulation. Implementation aspects of artificial intelligence languages and expert shell systems will be surveyed and possible limitations pointed out. The chapter concludes with descriptions of analog design systems introducing new features such as BLADES, AIDE2 and PROSAIC.

## 1 INTRODUCTION

In the VLSI design area there are specific design phases or circuits which are the most time consuming. Generally those areas are large integrated digital systems and full custom realizations, which also include analog functions. The fast development of semiconductor technologies makes it possible to realize more extreme specifications like speed [1-3]. It means that digital systems are becoming faster and more analog in nature.

Also new type of analog performance will be needed within integration of sensors and high speed signal processing.

The capability of using efficiently the technology maturing in research laboratories needs new short design cycles also for realizing analog functions. New VLSI design tools are today reaching the goal to shorten the design time of large digital systems and the compilation methods for integrated analog functions have been introduced.

New implementations of AI languages as well as the development of knowledge based systems are making possible to add new features also in the analog design environment. As a result of the maturing AI technology expert systems have been introduced also as a part of analog design environment. Even if the development within AI area has been fast the AI based analog IC design systems are still under research [4-7].

# 2 THE ANALOG DESIGN ENVIRONMENT

There are similarities but also differences within digital and analog design area. The hierarchy seems to be one of the obvious common features. However the abstraction levels in analog design are more unclear compared to the widely accepted behavioral, functional, register transfer, gate, circuit and transistor levels for digital systems. Within analog or analog-digital design similar abstraction levels like behavioral, functional, circuit and transistor level can be recognized but the boundaries between these levels are not sharp. This basic difference appears in design style as well as in synthesizing methods, Figure 1.

The differences of the design style requires systematic and heuristic, partially experimental decision making during the design. In analog and analog-digital design the last two aspects are emphasized and some demonstration design systems introduced [8-10].

In digital design a pure top down approach and abstraction level specific synthesizing tools are forming the basis for the automatic compilation. In analog design there does not exist generally accepted synthesizing methods because of the unclear abstraction levels. Because more than one abstraction level refers to the technology information, mixed top down and bottom up approaches must be utilized in analog synthesis e. g. iteration between the technology information and different abstraction levels is necessary.

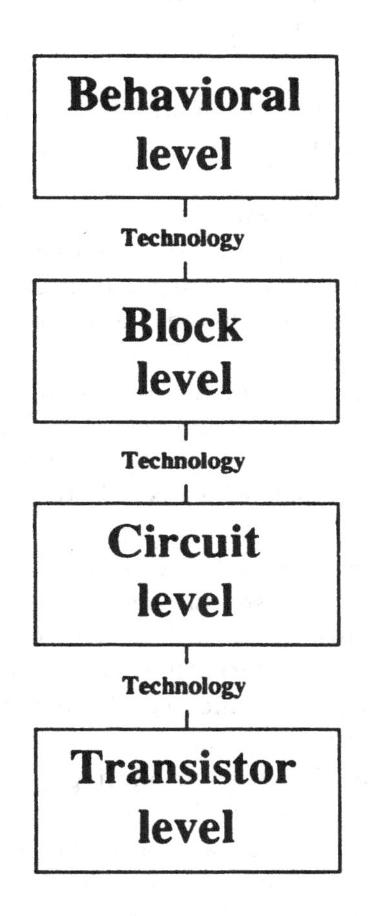


Figure 1: Schematic picture of the iterative nature of the analog and digital design.

# 3 SYNTHESIZING METHODS FOR THE ANALOG DESIGN

At the moment four synthesizing methods for integrated analog functions have been introduced: layout oriented, parameterized, systematic and knowledge based approach.

Layout oriented - semi-custom analog - approaches are relying on the previously characterized, limited number of circuit elements [8,9]. These methods are normally effectively using automatic placing and routing and the iteration between the technology and different abstraction levels is minimal.

Parameterized approach uses fixed or partially fixed topology or layout of the circuit and it can well be suited for regular topologies like A/D-converters and SC- filters [10]. This approach uses also the commercial CONCORDE compiler which is able to produce layouts for several analog functions.

Systematic synthesis approach follows similar hierarchy, selection and translation mechanisms used also in digital design [7,11]. Circuit topologies are selected from fixed alternatives not by constructing element by element. The fixed alternatives for circuit topologies are specified hierarchically and performance specifications are translated from the higher to lower level hierarchy.

Knowledge-based approaches are offering possibility to make synthesis partially based on exact and experimental information. A rule based system is able to build up analog blocks like operational amplifiers stage by stage [12]. Separated expert systems dedicated to different sub- blocks are demonstrated under manager expert to be able to synthesize higher level analog circuits [13]. Also separate specification experts can choose the most suitable circuit topology and combine the parameterized and knowledge based approach [14].

Not depending on the synthesizing method chosen also some optimization based on accurate enough models are needed within automatic analog design. There does exist several suitable optimization methods applicable in the analog design area but the problem is the accuracy of the modeling versus computing time needed during optimization [15-17]. The key questions of effective automatic design of analog functions is how the iterative, partially experimental design style and accurate enough optimization are combined with the synthesis.

# 4 GENERAL COMPONENTS OF A KNOWLEDGE BASED EXPERT SYSTEM

Widely seen an expert system can be any system which can do reasoning, but by adding an ideal, possibly open component - a knowledge base - to the reasoning mechanism it is possible to determine the common characteristics of a knowledge based expert system [18].

The structure of knowledge based systems varies but most of them have three parts: a working memory part, a rule memory part and the inference mechanism part or the rule interpreter. Normally the working memory has a collection of attribute-value pairs which describe the current situation. The rule memory part consists of different conditional type statements which operate on elements stored in the working memory. In deductive systems the rule memory is normally described with a high level type language. There also exist open, inductive systems which are producing rules based on the given examples probable realized and verified before.

The most important part of the knowledge based system is the inference mechanism. This part matches the working memory elements against the rule memory to decide which rule apply to the given situation. The selection process of rules can be data driven, goal driven or a combination of data and goal driven, Figure 2.

In most knowledge based applications the expert knowledge and the inference mechanisms are separated. Normally the decision making can be followed by the user as well as by why additional information was needed during the reasoning. The querying the user during the decision making can be connected both in forward and backward chaining mechanisms. The knowledge based approach has been applied to many fields especially in medicine and process control, where the interaction with the experimental and the on-line exact information is necessary.

The experimental nature of the analog and mixed design can effectively utilize induction in the generation of the reasoning rules. It is also obvious that the combination, data and goal driven, hybrid mechanism in reasoning is the most promising for the iterative, analog design. This is based on the fact that the rule memory part is always imperfect e.g. caused by the rapid development of the semiconductor technology. Some of the novel expert system shells can use this hybrid technique.

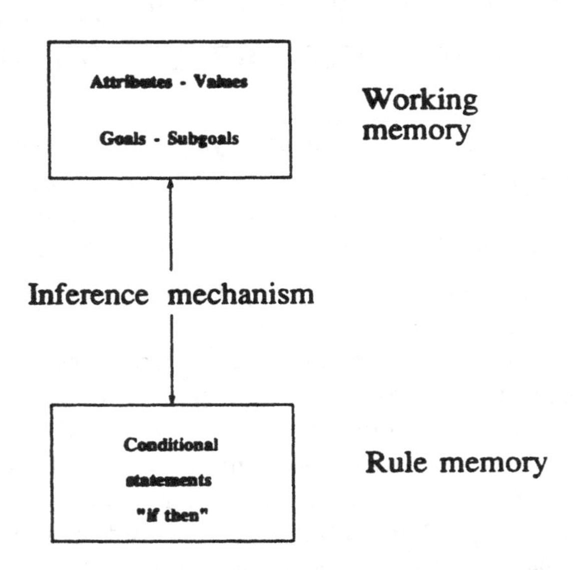


Figure 2: A general structure of the knowledge based system.

The present knowledge based systems are realized with different languages like C or AI languages like Lisp or Prolog. Several empty expert system shells have been implemented in the different computing environments [18], Table 1.

Obviously the front end of the analog and mixed design area can not be based purely on reasoning but requires some parts of synthesis including optimization and also simulation for the verification. The reasoning can be used partially as a synthesizing tool for iterative choosing and hierarchical building the circuit topologies based on the technology and specification information. The need for optimization and simulation as a part of the synthesis is exceeding the limits for mathematical problem solving of the present knowledge based systems. This requires possibilities for bidirectional communication between the knowledge based system and programs outside like optimizers and simulators.

Table 1: Comparison of four expert system shells and implementations of two AI languages for a front end of analog design.

| System/ language | Written in/pro- ducing | Reasoning | Inductive/ deductive | I/O to the user/ to programs outside | Hard- ware |
|---|---|---|---|---|---|
| KDS | Assembler/ none | Hybrid | Inductive | Menus/ assembler | IBM/PC AT |
| MUSE | PopTalk | Hybrid | Deductive | Textual, menus/ C oriented | SUN |
| NEXPERT OBJECT | "C" | Hybrid | Both | Graphical, menus/C-oriented, callable | VAX, SUN, MACII IBM/PC AT |
| RULE-MASTER2 | "C"/"C", Fortran | Hybrid | Both | Textual, menus/C Fortran | VAX, SUN, Apollo HP9000 IBM/PC XT, AT |
| GC Lisp | Language itself | Hybrid | Both | Textual/ via system calls | IBM/PC AT |
| Prolog2 | Language itself | Hybrid | Both | Textual/ C, Pascal, via system calls | IBM/PC AT |

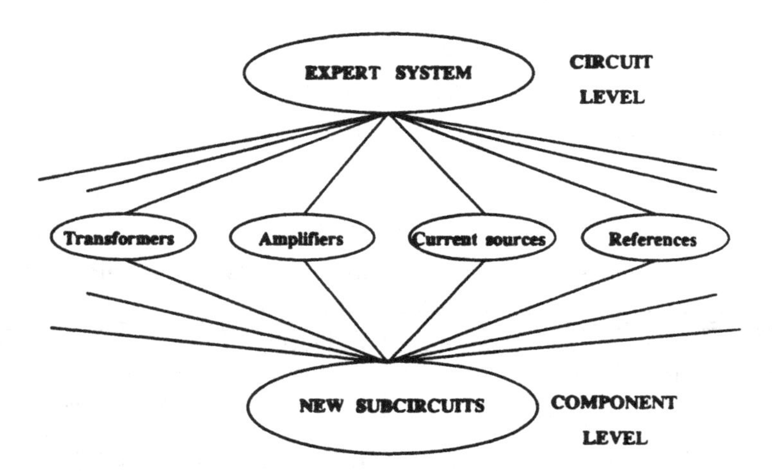


Figure 3: Schematic picture of a knowledge based BLADES- system, which is intended to synthesis basic analog building blocks [13].

# 5 EXPERIMENTAL SYSTEMS FOR ANALOG DESIGN

The oldest knowledge based systems in integrated circuit design is PALLADIO [18] and one of the first demonstrations with using knowledge based system for analyzing circuits was EL [19]. Most of the demonstrated systems in integrated circuit design area are directed to digital design but some demonstrations in analog design area have also been made, Figure 3.

The BLADES-system works as a separated front end and divides the problem according to the specifications into subproblems. The system works as an experienced circuit builder and consultant. It uses ready simulators like ADVICE and ANSYS for fine tuning the circuit. The control structure of the BLADES system is presented in Figure 4.

The BLADES-system presented in Figures 3 and 4 has some typical features of knowledge based systems for analog and mixed design area: dedicated working area, limited number of architectures available and need for bidirectional communication with the simulation programs outside. The system is mainly used as a consultant aid for the designer emulating the way to design analog circuits.

The decomposing nature of knowledge based analog system is presented well in the system called PROSAIC which uses hierarchy by de-

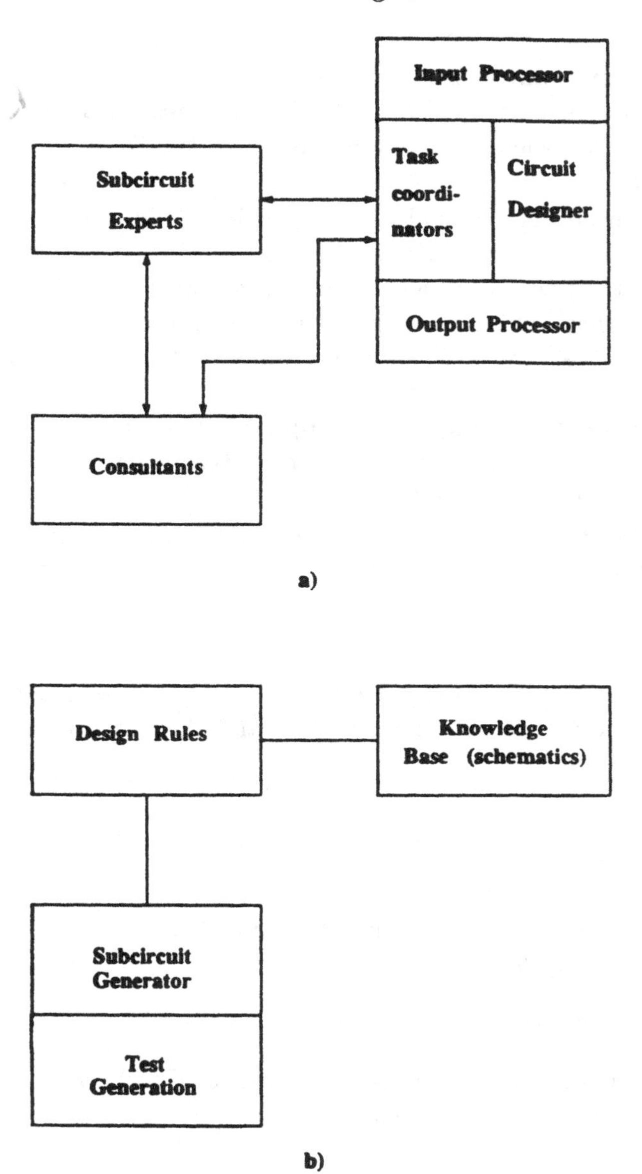


Figure 4: a) The control structure and b) the generation mechanism of a subcircuit of the BLADES-system [13].

composing circuits into smaller blocks, stages. It also selects the topology of different sub- blocks from the earlier designed alternatives described in the knowledge base, and produces the SPICE input list as a result, Figure 5.

The decomposing principle presented in PROSAIC is an approach to combine the systematic synthesis and knowledge based approach. It is also possible to combine the parameterized and knowledge based synthesis approaches. This idea requires the availability of an optimization package which can make the dimensioning according to the specifications.

An experimental parameterized approach is well presented in AIDE2 [10]. The program uses soft macros parameterizing the blocks according to the specifications, Figure 6.

With combining the a knowledge based system to the parameterized approach an intelligent design front end could be used for:

- Specifying
- Selection of different topologies according to the specifications and the experience from the technology and the different circuit topologies
- Interfaces to
    1. the user
    2. the verifying simulation environment
    3. the optimizer(s) if available
    4. the different technologies
- Decision making concerning acceptance of the optimized and simulated results versus specifications

Interface to the measured data or even to the measurements can be seen as a future trend.

It is essential that in the knowledge based parameterized approach the knowledge based system performs only those tasks to which it is suited well. The selection of the topology can be made mainly based on the experimental results and the characteristic performance of the technology.

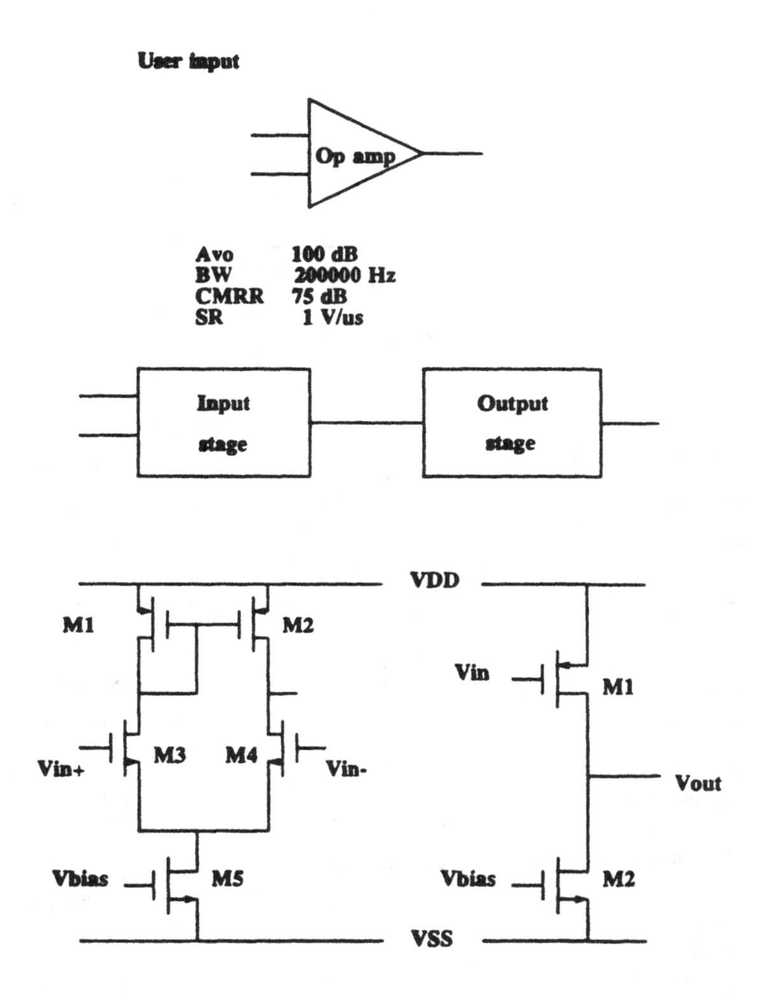


Figure 5: The decomposition principle of the PROSAIC program [12].

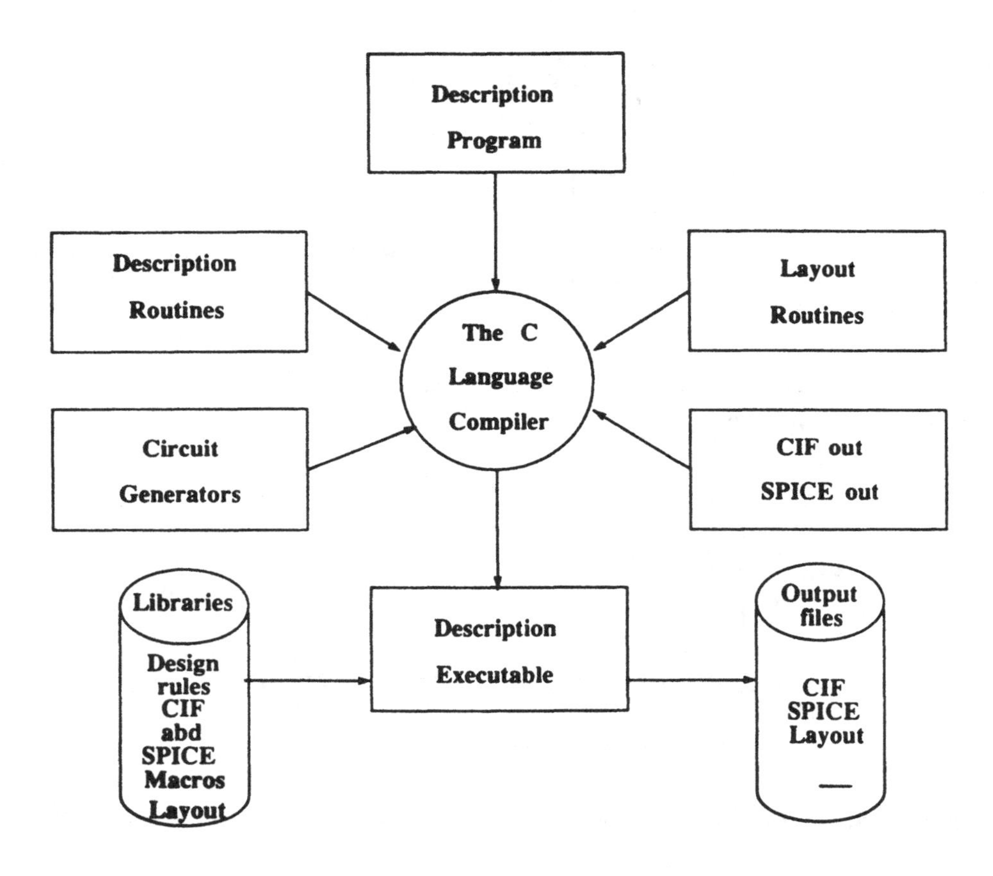


Figure 6: Structure of AIDE2, a parameterized, approach for synthesizing and verifying SC building blocks [10].

# 6 CONCLUSIONS

The automation of the analog design seems to be driven by the development of synthesizing and optimization methods for analog functions and AI techniques directed to the open, inductive knowledge based systems, which are using hybrid reasoning mechanism. Because synthesizing and optimization are a specific part of the automatic design of analog functions, the knowledge based systems must either include effective mathematical operations or smoothly interface with the outside optimization and simulation programs preferably in a workstation environment.

## References


[1] R. P. Jindal, Gigahertz-Band High-Gain Low-Noise AGC Amplifiers in Fine-Line NMOS, IEEE Journal of Solid State Circuits, Vol. SC-22, 1987, pp. 1031 - 1040.

[2] R. E. Johnston, A. Yie-Cheong Tang, A 600 MHz P- Channel JFET Compatible with an 8 GHz Bipolar Process, IEEE Custom Integrated Circuits Conference, Proceedings 1987, pp. 341 - 344.

[3] S. J. Nightingale, M. A. G. Upton, U. K. Mishra, S. C. Palmateer and P. M. Smith, A 30 GHz Monolithic Single Balanced Mixer with Integrated Dipole Receiving Element, IEEE Microwave and Millimeter-Wave Monolithic Circuits Symposium 1985, Digest of Papers, pp. 74 - 75.

[4] H. Yaghutiel, A. Sangiovanni-Vincentelli and P. R. Gray, A Methodology for Automated Layout of Switched- Capacitor Filters. IEEE International Conference on Computer Aided Design, ICCAD-86, pp. 444 - 447.

[5] M. Kliment, A. Matinez, S. Dholakia and M. Yee, Compilation Approach for Standard Cell Library Generation. IEEE International Conference on Computer-Aided Design, ICCAD-86, pp. 531 - 534.

[6] J. W. Fattaruso, R. G. Meyer, MOS Analog Function Synthesis. IEEE Journal of Solid-State Circuits, Vol. SC-22, No. 6, 1987, pp. 1056 - 1063.

[7] R. Harjani, R. A. Rutenbar and L. R. Carley, A Prototype Framework for Knowledge-Based Analog Circuit Synthesis. IEEE Design Automation Conference, Proceedings 1987, pp. 42 - 49.

[8] G. I. Serhan, Automated Design of Analog LSI, IEEE Custom Integrated Circuits Conference, 1985, Proceedings, pp. 79 - 82.

[9] G. H. Allwine, Design of Analog Data Acquisition System Using Silicon Compiler Techniques, IEEE Custom Integrated Circuits Conference, 1985, Proceedings, pp. 495 - 497.

[10] P. E. Allen and E. R. Macaluso, AIDE2: An Automated Analog IC Design System, IEEE Custom Integrated Circuits Conference, 1985, Proceedings, pp. 498 - 501.

[11] C. D. Kimble, A. E. Dunlop, G. F. Gross, V. L. Hein, M. Y. Luong, K. J. Stern, E. J. Swansson, Autorouted Analog VLSI, IEEE Custom Integrated Circuits Conference, 1985, Proceedings, pp. 72 - 78.

[12] R. J. Bowman, D. J. Lane, A Knowledge-Based System for Analog Integrated Circuit Design, Proceedings of IEEE International Conference on Computer-Aided Design, 1985, pp. 210 - 212.

[13] F. M. El-Turky and R. A. Nordin, BLADES: An Expert System for Analog Circuit Design, Proceedings of IEEE International Symposium on Circuits and Systems, 1986, pp. 552 - 555.

[14] H. Pohjonen, An Expert System As A Front End In Analog Compilation, NORSILC/NORCHIP seminar 1987, Trondheim, Norway, 7 p.

[15] K. Madeson, O. Nielson, H. Schaer-Jacobsen and L. Thrante, Efficient Minimax Design of Networks Without Using Derivatives, IEEE Transactions on Microwave Theory and Techniques, Vol. MTT-23, No.10, October 1975, pp. 803 - 809.

[16] P. Heikkil, CMOS OP AMP Dimensioning Using Multiphase Optimization, NORSILC/NORCHIP seminar 1987, Trondheim, Norway, 6 p.

[17] G.L. Tan, S.-W. Pan, W. H. Ku and A. J. Shey, ADIC-2.c: A General Purpose Optimization Program Suitable for Integrated Circuit Design Applications Using Pseudo Objective Substitution Method (POSM), IEEE Transactions on Computer-Aided Design, Vol. 7, No. 11, November 1988, pp. 1150 - 1163.

[18] T.J. Kowalski, An Artificial Intelligence Approach to VLSI Design, Kluwer Academic Publishers 1985, 226p.

# CHAPTER 3

# AUTOMATED DESIGN OF OPERATIONAL AMPLIFIERS: A CASE STUDY

L. Richard Carley
Department of Electrical and Computer Engineering
Carnegie Mellon University
Pittsburgh, Pennsylvania 15213
U.S.A.

## 1. INTRODUCTION

As the number of devices which can be included on a single integrated circuit (IC) increases, the trend is to implement application specific systems in one, or a few, ICs (ASICs). The fraction of the total number of IC designs which are developed for a single application has been steadily increasing in recent years. Many board-level systems have begun to migrate toward single chip implementations, and many of these applications require analog circuits as well as digital ones. In order to be economically competitive, the time between the definition of a profitable application and completion of the design of a working ASIC chip must be minimized. Therefore, the use of computer-aided design (CAD) techniques is an absolute necessity for ASIC design. While sophisticated CAD techniques have been developed to automate the design of digital circuitry, limited progress has been made toward automating the analog circuit design task in general. For example, even for an ASIC design on which 90% of the devices are used in digital circuitry and only 10% of the devices are used in analog circuitry, the analog circuit design task might well be the limiting factor in the overall design time. Because of the sensitive economic environment for ASIC designs, any delay is extremely expensive due to the loss of potential sales. Therefore, there is strong economic need for improved CAD tools to support analog circuit design. In this chapter, a general framework for the synthesis of sized transistor schematics from performance and process specifications developed at Carnegie Mellon University will be described. This work is the result of collaborative efforts between Professor Rob Rutenbar, also with the Department of Electrical and Computer Engineering at CMU, myself, and many of our graduate students, that began in 1985.

Some ideas from digital design methodologies, such as standard cell libraries and module generators, have recently been applied to analog design tasks. However, such techniques usually have severe drawbacks; e.g., analog cell libraries allow the designer to make only crude tradeoffs among performance specifications, and they become obsolete rapidly in the face of technological evolution. Custom analog circuits are still designed, largely by hand, by experts intimately familiar with nuances of the target application and the IC fabrication process. Analog design is commonly perceived to be one of the most knowledge-intensive of IC design tasks: the techniques needed to build good analog circuits are not written down and codified, but exist solely in the form of the expertise of individual designers.

The work of the human analog circuit designer can be arbitrarily divided into three phases.

- creating analytical expressions ("equations") for the behavior of new circuit topologies, and creating appropriately simplified circuit and device models
- using analytical expressions to guide the selection of both an appropriate combination of circuit topologies and a set of device parameters which together will cause the whole system to meet its performance specifications
- designing the mask geometries from the sized transistor-level schematics

Issues pertinent to applying computer-aided techniques to the first phase will be discussed in Chapter 4, and the issues pertinent to the last phase are beyond the scope of this book (for more information see [1]). This chapter focuses on one approach to applying computer-aided techniques to the second phase of the analog design process as divided above: using analytical expressions to develop sized transistor-level schematics to meet specific performance specifications.

This chapter is organized as follows. The first section will discuss general approaches to the analog design synthesis problem and give an in depth description of the specific class of synthesis approaches that encompasses the one presented in this chapter: equation-based design. The following section will describe the basic structure of the synthesis framework to be presented, OASYS, and explain how it is motivated by traditional analog design approaches. In order to test the feasibility of this organization for analog circuit synthesis, a prototype system that designs operational amplifiers has been developed. The architecture and implementation of the prototype version of OASYS, will be described. The critical role of hierarchy in analog circuits will be presented and, methods to structure and exploit analog design knowledge will be described. Examples of how analog design knowledge is codified within OASYS are included. The following section will describe the domain in which OASYS operates, and will present an evaluation of the performance of OASYS and its overall capabilities for synthesizing analog circuit modules. Automatically synthesized circuits that OASYS has produced will be presented. The following section illustrates the capabilities of OASYS to explore the analog circuit design space, and the importance of this capability. The final section presents some concluding remarks.

## 2. BACKGROUND: EQUATION-BASED DESIGN

Some ideas from digital design methodologies, such as standard cell libraries and module generators, have recently been applied to analog design tasks. However, such techniques usually have several drawbacks, e.g., libraries allow the designer to make only crude tradeoffs among performance specifications, and they become obsolete rapidly in the face of technological evolution. Since many of these existing approaches have been discussed in Chapters 1 and 2, they will only be discussed briefly here.

One approach that shows the most direct influence of digital design ideas is the analog standard cell library. This approach provides a rapid path to silicon for analog functions already designed to the level of primitive devices and the cell layout can be statically stored [2, 3, 4, 5]. Standard cells provide more density, and placement and routing tools that accommodate the sensitive electrical characteristics of these devices can help lay these out [6]. Unfortunately, this approach offers little flexibility, since designers are limited to a library of fixed circuit blocks. The nature of the performance specifications of an analog block, quite different from those of a digital block, make storing sufficient designs to span a significant portional of the designable space impossible. For example, the operational amplifier designer described later in this chapter takes 18 performance specifications each of which is a continuous variable; thus, the design space is 18 dimensional. To simply make 3 values (low, medium, and high) available for each performance specification would require a library with almost **400 Million** operational amplifier designs. This approach tracks technology changes rather poorly as well, since a change in the process typically requires redesign and layout of all of the library cells.

A higher-level approach is the use of analog module generators. These fix some portion of a circuit's topology and parameterize the remainder into a limited number of choices. The result is a simple algorithmic generator for a particular circuit topology. Regular structures, such as switched capacitor filters, are particularly amenable to this approach [7, 8]. Some systems, such as AIDE2 [9], use both approaches incorporating standard cells as subunits of a module generator; experiences with a successive-approximation A/D converter generator designed in this style appear in [10]. Other compiler systems, such Seattle Silicon Technologies' CONCORDE$^{tm}$, include a set of analog module generators based on parameterized custom layouts for common analog functions [11]. While this is a step in the right direction, these limited-choice module generation approaches still leave much to be desired: frequently most of the important design choices are fixed; there may be many sets of reasonable design specifications which cannot be satisfied by the available parameter choices; the designer may only be permitted to make crude trade-offs among performance specifications; and substantial modifications to the generators may be necessary in the face of technological evolution.

A practical analog synthesis tool must deal with all these concerns; i.e., unlike digital tools, it must handle performance parameters that constrain continuous quantities (e.g., voltages or currents), and it must adapt rapidly (and automatically) to changes in the fabrication process. In order to achieve these goals, it is necessary to take an entirely different approach to generating a sized transistor-level schematic from performance specifications for analog circuit modules. As already indicated in Chapter 1, this new approach is based on an attempt to imitate the methodology of the experienced human analog circuit designer.

Recently, several tools which employ this new, equation-based, approach to automatically generating a sized transistor-level schematic from performance specifications for analog circuit modules have begun to appear. Notably, these include the following tools: IDAC [12, 13], OASYS [14, 15], which is the subject of this chapter, OPASYN [16, 17], An_Com [18], and CAMP [19]. These design knowledge-based tools represent the first successful approaches to automating analog module design that generalize to a broad range of analog circuits and adapt relatively easily to changes in the fabrication process. Note, for these more general analog module generators, each topology that can be designed represents an infinite number of possible designs because the selection of device sizes within a topology varies almost continuously with the required performance specifications. The analog circuit classes to which this approach has already been successfully applied extend from operational amplifiers and comparators to band-gap voltage references and delta-sigma modulation type oversampling analog-to-digital (A/D) converters.

At their most fundamental level, these new approaches are all based upon imitating the human analog circuit designer. Human analog circuit designers frequently develop designs by solving simplified algebraic equations, often on the back of cocktail napkins. Their experience guides their choices of simplifications. Module designs which are promising are then simulated in order to determine their performance with more accuracy. In some cases, analog circuit designers have structured these sets of equations that described the approximate relationships between circuit behavior and various design variables, with a simple form of flow-chart, to represent the process of designing analog circuit modules; e.g., band-gap references [20].

In order to automate this process, it is necessary to extract and understand the design knowledge base used by the human analog circuit designer. The design knowledge base is represented in the following manner. A circuit topology is selected, simplified models for each active device are chosen, and a set of analytical equations which describe the behavior of the circuit are derived by the expert human designer, who can make knowledgeable simplifications. Computers can also aid in this part of the design process, as will be discussed in Chapter 4. The topology itself is part of the design knowledge base. Using this approach, the computer can never "invent" a new topology, it can only automate the selection of topologies from the overall library of topologies and the choice of component values for the selected topology. The analysis equations express the approximate performance of the circuit in terms of the process parameters and the component values.

In some cases it is possible to directly solve the analysis equations for the component values in terms of the performance and process specifications. However, frequently there are not enough equations to constrain all of the component values; the human designer removes these extra degrees of freedom by attempting to maximize or minimize overall system specifications (e.g., active area and total power consumption). Since the analysis equations are frequently nonlinear, it is often difficult to solve them directly in order to determine the device sizes and operating points. All of the equation-based approaches mentioned above start with analysis equations; however, they differ in how the component values are determined from performance specifications, and these differences have a substantial impact on their capabilities.

In this chapter, the discussion will focus on the OASYS synthesis system developed at Carnegie Mellon University. OASYS does not solve the analysis equations itself. Instead, it requires the expert human designer to provide it with a set of design equations in addition to the analysis equations. These design equations are formed by solving the analytical equations for the circuit component values in terms of the desired performance values and the process parameters. Thus, the design equations consist of an ordered set of equations, each of which expresses an as yet undetermined component value or intermediate variable in terms of performance parameters, process parameters, and previously determined variables. It relies upon a fixed point iteration technique, to be described in more detail in the next section, to solve nonlinear equations. The next section will describe the process by which OASYS manipulates the design equations and analysis equations in order to carry out the design process.

In some cases, these ordered sets of equations have even been put in the form of large computer programs that automatically design all or part of a particular analog circuit module using a particular circuit topology; e.g., [13, 21]. However, the key feature of the OASYS framework is that it formalizes the basic approach: its goal is to separate the definition of a specific language in which design knowledge can be written from the interpretter that operationalizes that knowledge. The importance of explicitly formalizing the process is clear. As these tools mature large libraries of design knowledge will be created, and analog circuit designers will find it to their advantage to leverage their design time by exploiting this computer store of design knowledge. This design knowledge will only be easily reused if it is not spread out within a large program, most of which is required for "turning the crank".

One final note, there are limitations on using the design knowledge-based approach to analog circuit synthesis. The process of creating the design knowledge base for a circuit topology is much more time consuming than adding a cell to an analog cell library that meets a single set of specifications. Therefore, creating a design knowledge base by hand will only be economical for frequently used analog modules. As will be discussed, this limitation is eased somewhat for design systems that employ hierarchy, like OASYS, since a high level module can be described in terms of a few medium level modules rather than many individual devices.

# 3. THE OASYS SYNTHESIS FRAMEWORK

In this section, the CMU analog synthesis framework, OASYS, will be presented. The specific goal of OASYS is to convert a behavioral circuit description into a specific structural description; i.e., the inputs to the OASYS synthesis framework are circuit performance specifications and process specifications, and the output is a sized circuit schematic.

Since OASYS is based upon analytical approximations that characterize the behavior of specific circuit topologies, the circuits designed by OASYS may not precisely meet all of the performance specifications. This is not a significant drawback to this approach for several reasons. First, the designs produced by

OASYS can be further refined by using conventional circuit optimization tools [22]. The prime goal of OASYS should be to choose the best topology and to choose devices sizes that are approximately correct. Even expert designed circuits frequently need fine tuning using a circuit simulator. Second, due to the wide variations in process parameters, there are always variations in the performance of analog circuits. In the current version of OASYS, an adjustable overdesign factor can be assigned to performance specifications in order to insure that the desired performance will be achieved in spite of inaccuracies in equations and variations in process parameters.

## 3.1. Why a Framework

The underlying goal of our research is to provide a general framework that can be used to automate analog circuit module design (see Fig. 3-1). Note, it is much more difficult to build an environment to support the design of a broad range of analog circuits than it is to write one program to design one circuit. However, as already stated, there are many important benefits to having a clean separation between the domain knowledge and the software which operationalizes the design task. As will be seen, our views of a framework are based upon a computational model of how circuits are designed by human expert designers; i.e., it relies heavily on domain knowledge.

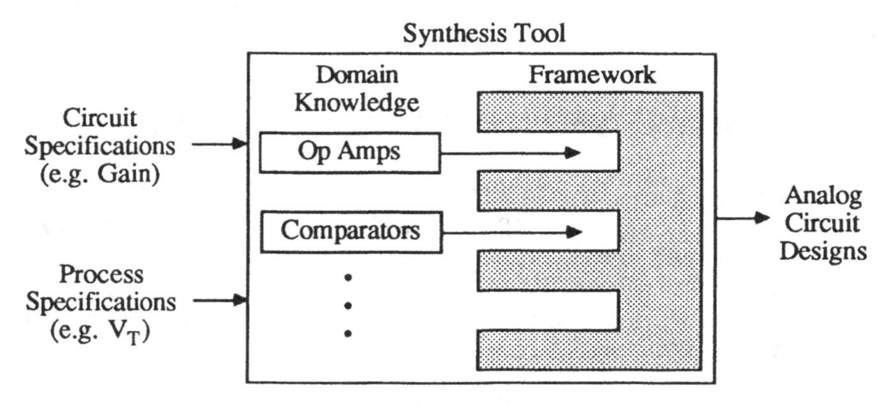


**Figure 3-1:** The OASYS Framework - Design Knowledge Versus Program Engine.

There are a number of important requirements for a general framework for analog circuit design:

- It must provide a uniform, structured view of synthesis.
- It must compartmentalized design knowledge, which greatly facilitates the addition of new knowledge and the modification of existing knowledge.
- Design knowledge must be reusable: it should be acquired once and available everywhere.
- It must be easy to extent design knowledge: ease to add new knowledge and easy to modify existing knowledge.

In the next section we will see how the OASYS framework achieves these goals.

## 3.2. Key Ideas

There are three key ideas that underlie the OASYS framework: that analog circuits do admit a hierarchical decomposition, that for any circuit design there exists a group of known block-level topologies - one of which must be selected, and that equation-based methods can be used to translate specifications from one level in the hierarchy to a lower level.

### 3.2.1. Hierarchy.

Unlike digital circuits, analog circuits do not generally have a universally agreed upon and unique hierarchical decomposition; e.g., some researchers view an op amp as made up of stage-1 and stage-2 while others decompose the op amp into differential pairs, current mirrors, etc. However, hierarchical decomposition does play a very important role in the way human analog circuit designers attack the design task (see Fig. 3-2). And, a natural method for determining a "better" hierarchical decomposition is to compare how much information must flow between levels of the hierarchy: the less information that flows, the "better" the choice. This is consistent with our desire that the use of hierarchy should decompose the problem as much as possible. The fundamental viewpoint embodied in the OASYS framework is that the hierarchical structure is necessary to decompose the design, creating many individual design tasks which are smaller and easier to implement.

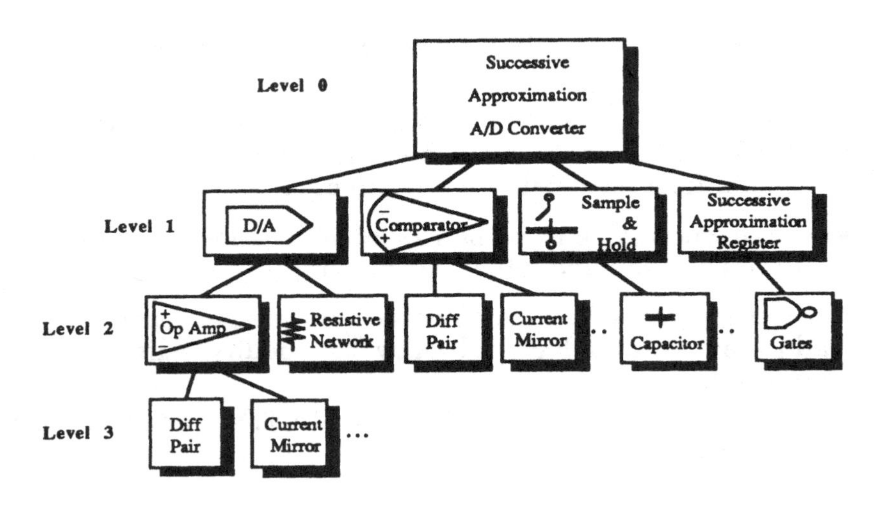


**Figure 3-2:** Example of Hierarchy in Analog Circuits.

The design process follows this hierarchical description from the top down to the leaf nodes. The finer the hierarchical decomposition, the fewer the number of variables which must be dealt with in any single step. Note, as will be discussed in more detail below, the complexity of the analog design task requires that rather than proceed unidirectionally from top to bottom, the actual design process requires negotiation up and down in the hierarchy in order to arrive at a design which can meet the necessary performance requirements. One disadvantage of hierarchy is that optimizing the choice of sub-block performance is more difficult in a hierarchical structure than in a flat transistor-level schematic.

Another important advantage of using a hierarchical decomposition is that a piece of the solution to one design problem may be reusable in another. For example, the current mirror design strategy developed as part of the 2-stage op amp design strategy was reused in 1-stage op amp design strategy. Also, since submodules may be incorporated into many different circuit topologies, but the domain knowledge associated with its design is stored in a single place, improvements in a sub-module's domain knowledge propagate automatically to all circuit topologies which incorporate that sub-module.

Employing hierarchy, even at the level of an op amp, is a

great benefit because each of the component sub-modules of an operational amplifier may be designed in several possible topologies. For example, OASYS currently supports only two block-level topologies for op amps; however, it can design 72 distinct practical device-level topologies. Both of the op amp topologies are constructed from sub-modules like differential pair or current mirror rather than being specific topologies of individual devices. This is an extremely important advantage of hierarchy because in many cases the reason automated design systems are not able to match the performance of expert human designers is that human designers are able to select from a wider range of topologies.

A second potential drawback of a hierarchical structure is that it is not possible to implement design tricks in which knowledge about choices made in one sub-module influence choices made in another sub-module. However, with careful choice of the hierarchy this limitation is typically not significant.

**3.2.2. Design Style Selection.**

A design style is an interconnection of abstract blocks. We choose this terminology to be consistent with the more evolved work on digital circuit synthesis. Because of the hierarchical decomposition, design styles are topologies of connected blocks, not transistors; i.e., choosing a design style is not choosing a transistor schematic. The OASYS system relies on the existence of mature design styles at each level of hierarchy for the analog circuit modules to be designed. The choice of the hierarchical decomposition and of the design styles at each level is an important part of the domain knowledge which must be coded into OASYS by an expert analog circuit designer.

Typically at any one level in the hierarchy, there are several design styles available to implement a particular function; e.g., we have two design styles for a current mirror, a standard current mirror and a cascoded current mirror. Normally, these different design styles provide the same functional behavior, but offer different performance tradeoffs. Frequently they employ different sub-blocks and/or sub-block topologies.

At any point in the design process, we must design a block with some desired functional behavior. In OASYS, we first select one of the available design styles to pursue based on the necessary performance specifications. In addition, since the selected design style might fail to meet our performance specifications, we must detect design styles that are incompatible with the specifications, and -- to the extent possible -- rank the compatible ones (see Fig. 3-3). For example, if the design specifications for an op amp require differential outputs, then an op amp design style with only a single ended output is incompatible.

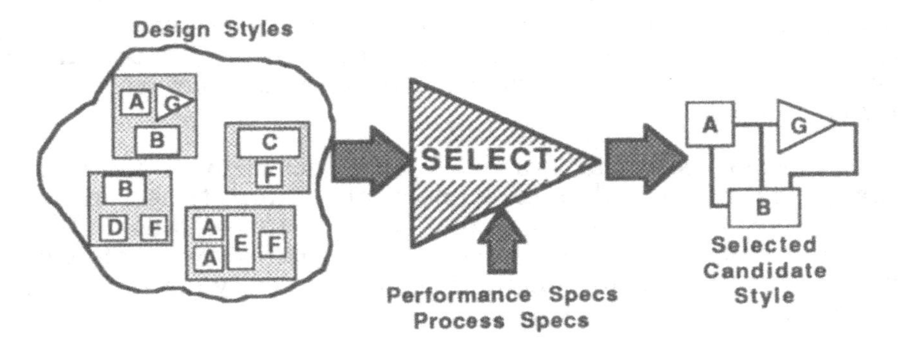


**Figure 3-3:** Example of Design Style Selection.

### 3.2.3. Translation.

Having selected an interconnection of sub-blocks, the next step in the design process is the determination of performance specifications for each of the sub-blocks that make up the selected design style. We define this process as translation of the performance specifications at the block level into a set of performance specifications at the sub-block level. The translation process must guarantee that if sub-blocks with the chosen performance specifications are connected in the manner indicated in the selected design style, then the block will meet its performance specifications (see Fig. 3-4).

In OASYS the translation process uses analytical models of the relationships between the performance specifications of the sub-blocks and the performance specifications of the block that are provided by expert human designers. The details of how the translation process proceeds will be examined in much greater detail in the next subsection.

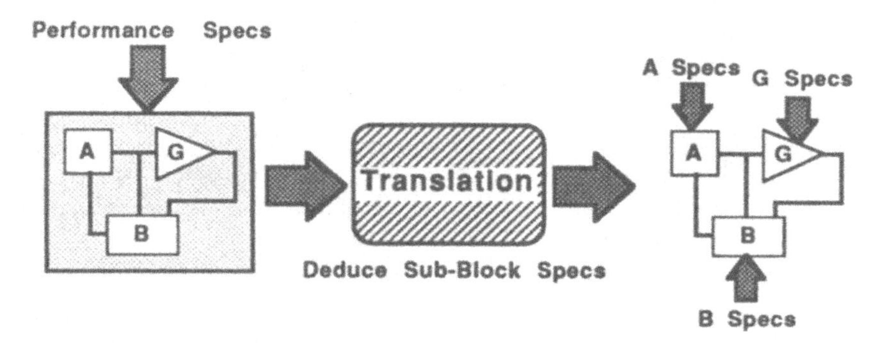


**Figure 3-4:** Example of Translation of Specifications.

### 3.2.4. Completing the Design.

A complete design consists of a hierarchical decomposition, and a set of design style selections and associated translations. That is, we start by selecting a design style for the block, which consists of an interconnection of sub-blocks. We then translate block specifications into sub-block specifications. Then, for each sub-block we select a design style for that sub-block, each of which consists of an interconnection of sub-sub-blocks. We then translate sub-block specifications into sub-sub-block specifications for every sub-sub-block. This process of selection and translation continues recursively until the sub-sub---blocks are primitive components; e.g., transistors, C's, R's (see Fig. 3-5) Only at this bottom level does translation correspond to device sizing. Note, since the analog hierarchy is not strict (uniform in depth along each branch): the component-level may be reached at different depths along different branches.

Hierarchy compartmentalizes the design process by establishing obvious boundaries between the selection and translation tasks at each level of the design process. This is one reason that hierarchy enhances the reusability of sub-blocks. Any part of the hierarchy is a complete design entity which can be incorporated into other designs or used independently. Finally, the hierarchical decomposition establishes a uniform design process at all level: selection followed by translation. This uniformity greatly facilitates the construction of a framework that implements the underlying activities of the design process at any level. The difference between different levels in the hierarchy is completely

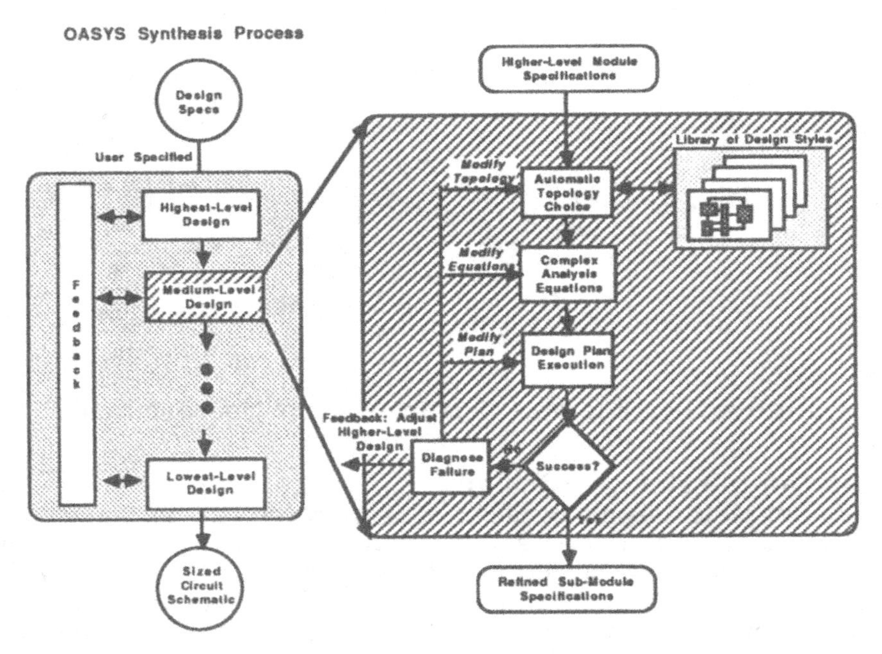


**Figure 3-5:** Block Diagram of OASYS Framework.

captured by differences in the domain knowledge: the framework's operation is identical at all levels.

## 3.3. Translation - An In Depth Look

In OASYS the translation process uses analytical models of the relationships between the performance specifications of the sub-blocks and the performance specifications of the block that are provided by expert human designers. We refer to this as part of the domain knowledge for this particular block. In this section, equation-based strategies for translation will be presented in detail. To make the ideas of hierarchy and translation more concrete, this section presents two extremely simplified examples of OASYS-style synthesis: one example at the transistor level and a second example at the op amp level. For the sake of brevity, pieces of the overall synthesis process have been selected to illustrate the process.

### 3.3.1. Current Mirror Design Example.

In the following example, we will use the current mirror design strategy to illustrate the process of translation, first for a very simple current mirror design style and then for a more complex current mirror design style.

Consider a simple MOS current mirror design strategy which accepts three performance specifications (as described in the previous op amp design strategy) and four process specifications:

1. Maximum Output Mirror Current, $I_o^{max}$
2. Minimum Output Resistance, $R_o^{min}$
3. Minimum Output Voltage, $V_{o-M}^{min}$
4. $k'$ for the process
5. $V_T$ for the process
6. $L_{min}$, the minimum allowable device length in the process.
7. $\lambda$ for the process, where the device output resistance is given by $(1/\lambda I_D)(L/L_{min})$

First consider the design style for a simple 1:1 ($I_{IN} = I_{OUT}$) current mirror as shown on the left in Fig. 3-6. The translation process for the simple current mirror is completely algorithmic under the assumption that the smallest possible device width $W$ and device length $L$ will be chosen in order to minimize area:

1. Choose $L$ to achieve desired mirror output resistance $R_o$, using the fact that $R_o = (1/\lambda I_D)(L/L_{min})$.

   Therefore: $L = L_{min} R_o^{min} \lambda I_o^{max}$

   If $L < L_{min}$ then $L = L_{min}$
2. Choose $W$ to achieve desired output voltage $V_o^{min}$, using the fact that $I_D = k' W(V_{GS}-V_T)^2/2L$.

   Therefore: $W = 2LI_o^{max}/k'(V_o^{min})^2$.
3. **Failure:** we could define a maximum dimension for $L$ and $W$ which if exceeded would cause us to classify this translation as a "failure"; i.e., the input

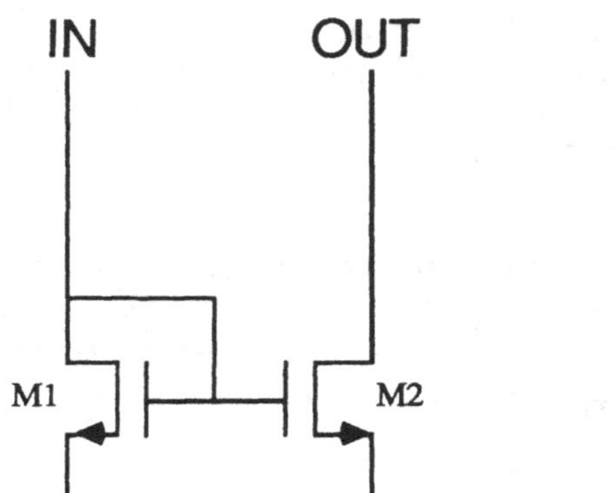


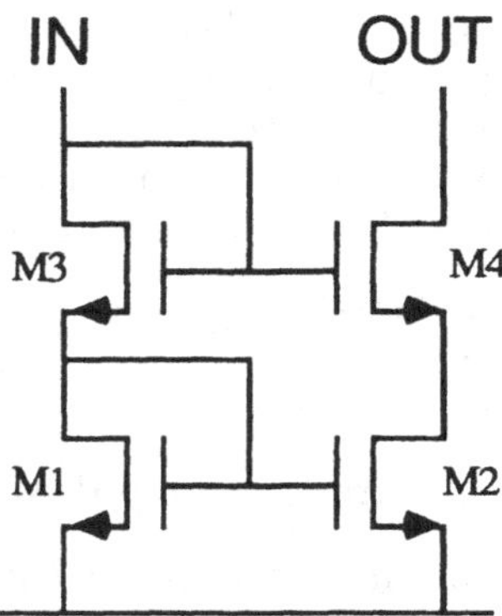


**Figure 3-6:** Simple and Cascoded Current Mirror Schematics for Example.

specifications could not be translated to a "reasonable" design.

Now consider the design style for a 1:1 cascoded current mirror as shown on the right in Fig. 3-6. Because many possible combinations of device sizes can achieve the same performance specifications, the translation process for the cascoded current mirror employs a heuristic in order to narrow the design space. In this example, we assume that the $W$ and $L$ of all four devices are equal. Note, many other heuristics are possible (and may be better); e.g., the length and width of two cascode devices could be set to their minimum size and the width of all four devices could be equal. However, the OASYS framework does not attempt to choose the best heuristic; rather, it simply automates the heuristic provided in the form of domain knowledge by the expert analog circuit designer. The translation process proceeds as follows:

1. Choose $W/L$ ratio to achieve desired output voltage $V_o^{min}$, using the following facts (and neglecting the body effect):

$$\Delta V_{GS} = (V_o^{min} - V_T)/2$$
$$I_D = k' W(\Delta V_{GS})^2/2L$$

Therefore: $W/L = 2I_o^{max}/k'(\Delta V_{GS})^2$.

2. Choose $L$ to achieve desired output resistance $R_o$, using the following facts (again neglecting the body effect):

   $$R_o = g_{m4} r_{o4} r_{o2}$$
   $$R_o = (L/\lambda I_D L_{min})^2 K'(W/L)\Delta V_{GS4}$$

   Therefore:

   $$L = L_{min} \lambda (R_o^{min} I_o^{max} (V_o^{min} - V_T)/4)^{1/2}.$$

3. If $L < L_{min}$ then $L = L_{min}$

4. Finally, choose $W$ by substituting this computed $L$ into the equation for $W/L$ given in step 1 above.

5. **Failure:** we could define a maximum dimension for $L$ and $W$ which if exceeded would cause us to classify this translation as a "failure"; i.e., the input specifications could not be translated to a "reasonable" design.

Note, in the actual OASYS implementation, the current mirror design strategies are more complicated than indicated here, since they must handle current mirror ratios other than 1:1, device-level issues such as the body effect and matching, circuit-level issues such as noise and bandwidth, and the usual problems of verifying the validity of device models over different ranges of operation.

**3.3.2. Op Amp Design Example.**

In this second example, we assume that a specific (highly simplified) block-level topology for an operational amplifier has been selected and we present a simplified strategy for the translation of its op amp specifications into sub-block specifications. As mentioned in the previous section, more complicated nonlinear analysis equations cannot be solved directly, and some means must be provided by the framework to cope with their solution. OASYS uses heuristic guesses for the values of selected variables in order to facilitate the translation process, and the values for these heuristic guesses are included as part of the domain knowledge. Unfortunately, in many cases these guesses are incorrect. Part of the domain knowledge, provided by the expert analog designer, is where to test the validity of each of these heuristic guesses and if they are in error, how to proceed with the translation process. This typically involves changing the invalid heuristic assumption and restarting the translation process from that point. One of the important components of the framework is that it stores the complete history of the state of the design at each step, which allows the translation process to be "unwound" to any desired point.

Remember that we are assuming that the selection process has already occurred, and a highly simplified OTA style has been selected (shown in Fig. 3-7). The following is a simplified

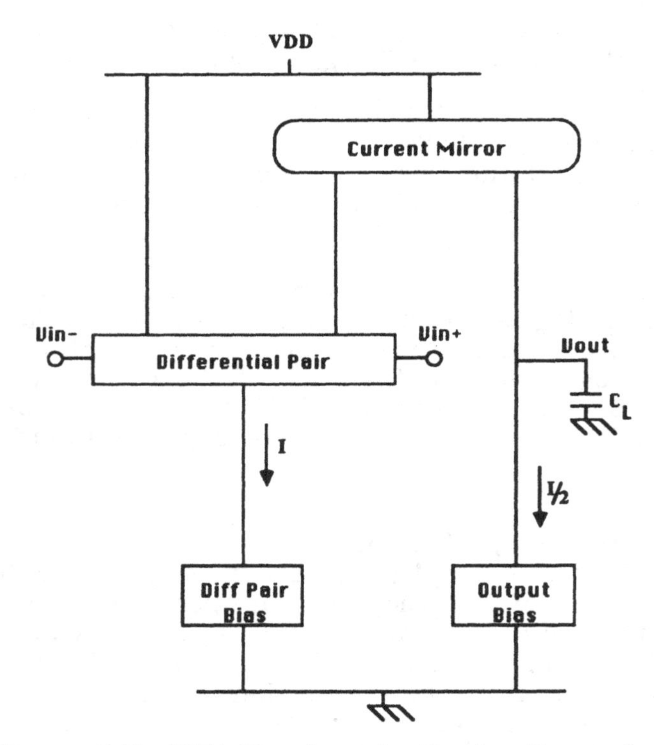


**Figure 3-7:** OTA Topology for Design Example.

strategy for translation of this op amp. Assume that the simple op amp design strategy accepts as input six performance specifications:

1. Slew Rate, *SR*

2. Load Capacitance, $C_L$
3. Unity Gain Crossover Frequency, $\omega_c$
4. DC Open Loop Gain, $A_o$
5. Maximum Output Voltage, $V_{o-OA}^{max}$ (The "OA" subscript denotes "opamp" and is used to avoid confusion when we need output voltages for sub-blocks later.)
6. Minimum Output Voltage, $V_{o-OA}^{min}$

Note, throughout this design strategy we will attempt to minimize power and active area to remove extra degrees of freedom in the design space. The design strategy is as follows:

1. Choose bias current $I$ to achieve desired slew rate $SR$, using the fact that $SR \leq (1/C_L)I$ and minimize power by choosing the smallest possible bias current.

   Therefore: $I = C_L SR$

2. Choose the transconductance of the differential-pair sub-block, $G_m$, to achieve the desired unity gain crossover frequency, $\omega_c$, using the fact that $\omega_c = G_m / C_L$. (Notice, $G_m$ is an *output* of this translation process, but is itself an *input* performance specification for the differential-pair sub-block.)

   Therefore: $G_m = C_L \omega_c$

3. Choose $\alpha$, where $r_{o-M} = \alpha r_{o-B}$, and $r_{o-M}$, $r_{o-B}$ are the output resistance of the current mirror and output bias current source sub-blocks, respectively. In this case there is insufficient information to make an optimal choice for $\alpha$, therefore, we make a *heuristic* assumption based on experience.

   For simplicity, we *assume*: $\alpha = 1$.

4. Choose $r_{o-M}$ and $r_{o-B}$ using the fact that $A_o = G_m (r_{o-M} \| r_{o-B})$.

Therefore: $r_{o-M} = \alpha r_{o-B} = (A_o \ / \ G_m) \ (1+\alpha)$.

5. Design the current mirror sub-block using the following input performance specifications:

   Maximum Output Mirror Current, $I_o^{max} = I$

   Minimum Output Resistance, $R_o^{min} = r_{o-M}$

   Minimum Output Voltage, $V_{o-M}^{min} = V_{DD} - V_{o-OA}^{max}$
   (As before, we use a subscript, "M" to denote parameters for the mirror.)

   **Failure Recovery:** If current mirror design fails (as already described) then increase $\alpha$ and restart at step 4.

6. Design the output bias current source, the differential-pair, and the differential-pair bias current source (these are similar in style to the previous steps; we omit details.).

7. Adjust $\alpha$ in order to decrease total active area of all modules -- note, only the Output Bias Current Source and the Current Mirror are affected by the choice of $\alpha$. Compute sensitivities of Mirror area, $A_m$, and Bias area $A_b$ with respect to $\alpha$. If $\partial A_m/\partial\alpha < -\partial A_b/\partial\alpha$ then increase $\alpha$, else decrease $\alpha$. Then restart at step 4.

Note that steps 5 and 7 in the design strategy use a simple form of fixed-point iteration; step 5 iterates to find a workable value for $\alpha$, while step 7 iterates to optimize area.

Although the method of adjusting $\alpha$ is not described in detail here, a variety of numerical methods, or heuristic techniques based on designer expertise, could be employed. For example, a valid heuristic might simply be to adjust $\alpha$ by some constant amount. The critical point to note is the role of hierarchy in making traditional "flat-circuit" optimization impossible here. Because we refine to blocks like mirrors, *not* transistors, it is not possible to develop a set of flat equations that relate overall performance to device sizes and bias currents, and to apply optimization techniques that manipulate device parameters. Indeed, in this hierarchy, we do not even know yet the device-level implementations of the sub-blocks.

Within the OASYS implementation of real op amps, design strategies are considerably more complicated than indicated by this simple illustration, and handle many additional parameters; e.g., noise, frequency dependence and phase margin. Also there are limits on the range of validity of the models used to generate these design strategies, hence the translation process must verify that the results produced are consistent with modeling assumptions, and correct the design if they are not.

### 3.3.3. Iteration Strategy.

In the OASYS framework, as indicated in the op amp design example, the solution of complex analysis equations is done by making heuristic assumptions and then doing fixed point iterations. The heuristic assumptions to make, and the tests to verify if they are valid are all provided by the expert analog designer as part of the domain knowledge for the block translation. In general, OASYS design strategies first try the "nearest" changes (see Fig. 3-8).

OASYS employs a hierarchical structure which decomposes the design of an op amp into the design of several sub-modules. One result of this decomposition is that fewer nonlinear design equations must be solved simultaneously. OASYS solves these sets of nonlinear equations by a combination of starting from design knowledge-based heuristic assumptions and a having a set of test and control strategies which iterates through the design equations to adjust the heuristics. Because it uses this iterative improvement strategy, the performance of op amps designed by OASYS is comparable to that of circuits designed by expert human designers using the same fabrication process and the same circuit topologies. The next section provides specific designs synthesized by OASYS and their performance.

## 4. AUTOMATED ANALOG MODULE GENERATION

An important application of the OASYS framework is to use the domain knowledge to synthesize sized transistor-level

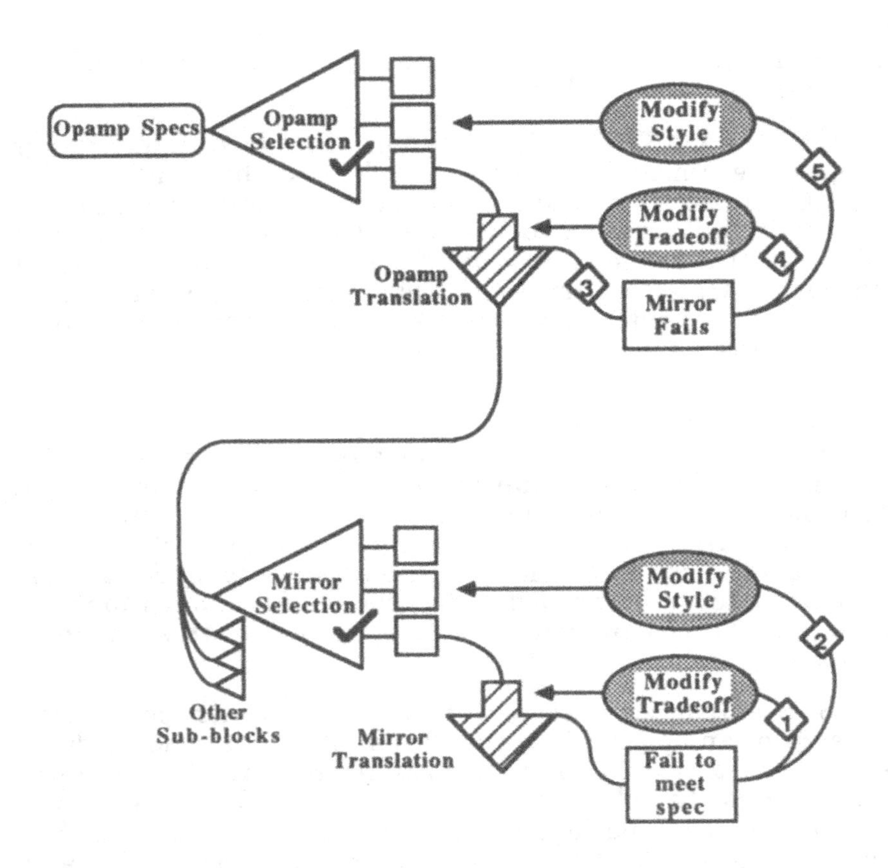


**Figure 3-8:** How the OASYS Framework Copes with Failure.

schematic diagrams from performance specifications and process specifications. In this section we will document the ability of OASYS to synthesize op amps. First, we will present specifics of the OASYS framework and the op amp domain knowledge base. Then, we will present examples of op amps that have been synthesized by OASYS.

## 4.1. Implementation and Applications

Operational amplifiers were chosen as an initial test domain for OASYS because they are ubiquitous components in many system-level designs, and because they appear to be the favored first target of other earlier synthesis approaches [23, 24, 25]. We restricted the domain knowledge to the design of CMOS op amps, not because of any limitation of the framework, but because we only have access to CMOS fabrication. The op amp hierarchy was chosen to explore two levels of design style selection and translation:

1. Select op amp design style.
2. Translate op amp specifications into specifications for sub-blocks such as current mirrors, differential pairs, etc.
3. Select design styles for each sub-block
4. Translate each sub-block specification into device interconnections and sizes.

OASYS currently incorporates two topologies for op amps: a standard Miller-compensated 2-stage style, and a 1-stage operational transconductance amplifier (OTA) style (see Fig. 4-1). Several lower-level blocks are also available, each currently in a few different styles: current mirrors, differential pairs, transconductance amplifiers, level shifters, compensation schemes, and bias blocks. Style selection and translation activities occur at each level of this hierarchy; the number of style choices is small at each level, but translation is still complex. Notice that even with these limited style choices, the hierarchical composition of styles yields several hundred different transistor-level op amp topologies, of which about 72 are practical circuits commonly seen in real applications.

Most of these design styles are not specific to a particular topology: they are based on their own independent design strategies, and are fully reusable as parts of other higher-level designs. The only exceptions are the feedback compensation block and the bias block which are tightly bound to a particular parent style. All of the other lower-level blocks can be used directly in any new design strategies, e.g., in a folded-cascode op amp, a voltage reference, a comparator, etc. For example, the 2-stage op amp style required 18 months to implement, primarily because it was the first op amp style. To build 2-stage op amps, we had first to implement design styles for all the required sub-blocks, such as mirrors, differential pairs, level shifters and so forth. In addition, the framework itself was evolving as this design knowledge was being captured. In comparison, after

| Block Types | Style | |
|---|---|---|
| Operational Amplifier | One-Stage | Two-Stage |
| Current Mirror | Simple | Cascoded |
| Current Source | Simple | Cascoded |
| Differential Pair | Simple | Cascoded |
| Transconductance Amplifier | Simple | Cascoded |
| Level Shifter | Trivial | Follower |
| Feedback Compensation | R-C | Buffer |
| Bias Voltage Network | General | |
| Total number of distinct practical transistor-level topologies = 72 | | |

**Figure 4-1:** Blocks and Associated Styles Currently Implemented in OASYS.

deducing the appropriate hierarchy and heuristics for the OTA op amp design style, a design strategy was implemented in roughly one month. This is because all the sub-blocks were in place, and the framework itself was more stable. Any subsequent improvements in the design strategy for a particular block expand the range of performance specifications that can be attained by any higher-level topologies of which this block was a part. For example, an improvement in the current mirror designer would extend the range of specifications achievable by all our op amp design strategies.

OASYS requires as input a description of the fabrication process, provided in a technology file (see Table 4-1) and a set of desired op amp performance specifications. Because OASYS uses design equations in which process parameters are variables, it adapts easily to different processes. Only when the process is modified to such an extent that new device models are needed must the OASYS design equations be modified, and even then only in the lowest-level design styles. This ability of equation-based design methods to "scale" with technology is extremely important, and is a result of forming a library of design methods rather than specific designs. Table 4-2 gives the performance parameters that OASYS can currently design to meet. From these inputs, OASYS produces a sized transistor-level circuit schematic in the form of an annotated SPICE file [26].

| Process Parameters |
|---|
| 1. Threshold Voltage, p and n (V) |
| 2. $K'$, p and n ($\mu A/V^2$) |
| 3. Process Minimum Width and Length ($\mu m$) |
| 4. Built-in Voltage, p and n (V) |
| 5. Minimum Drain Diffusion Width ($\mu m$) |
| 6. Supply Voltage, $V_{DD}$ and $V_{SS}$ (V) |
| 7. Oxide Thickness (nm) |
| 8. Mobility, p and n ($cm^2/V{-}s$) |
| 9. $C_{gso}$, p and n ($fF/\mu m^2$) |
| 10. $C_{gdo}$, p and n ($fF/\mu m$) |
| 11. $C_{gb}$, p and n ($fF/\mu m$) |
| 12. $C_{jsw0}$, p and n ($fF/\mu m$) |
| 13. $C_{j0}$, p and n ($fF/\mu m^2$) |
| 14. Temperature (K) |
| 15. Flicker Noise Coefficient ($V^2 m^2 rad/s$) |
| 16. Channel Length Modulation, p and n, $f_0$ ($V^{-1}$), $f_1$ ($\mu m/V$), $\lambda$ ($V^{-1}$), p and n for $\lambda = f_0 + f_1/Length$ |

**Table 4-1:** Process Parameters Currently Used by the OASYS Framework.

The version of OASYS discussed in this paper, OASYS 0.3, comprises about 11,000 lines of Franz LISP running under UNIX[1] on a VAXstation[2] II/GPX. Details about the evolution of this implementation appear in [14] and [27].

[1]UNIX is a trademark of AT&T Bell Laboratories.

[2]VAXstation II/GPX is a trademark of the Digital Equipment Corp.

| Performance Parameters |
|---|
| 1. Voltage Gain (dB) |
| 2. Slew Rate (V/μs) |
| 3. $C_{load}$ (pF) |
| 4. Power Dissipation (mW) |
| 5. Unity Gain Frequency (MHz) |
| 6. Phase Margin (degrees) |
| 7. Common Mode Rejection Ratio (dB) |
| 8. Power Supply Rejection Ratio (dB) |
| 10. Offset Voltage (mV) |
| 11. Active Area ($\mu m^2$) |
| 12. Settling Time (μs) |
| 13. Input Common Mode Range (V) |
| 14. Output Range (V) |
| 15. Input Referred Flicker Noise Voltage ($nV/\sqrt{Hz}$) at Specified Flicker Noise Frequency (Hz) |
| 16. Broadband Noise Voltage ($nV/\sqrt{Hz}$) |

**Table 4-2:** Performance Specifications Currently Used by the OASYS Framework.

## 4.2. Conventional Synthesis

In this section, we present a number of OASYS designs that use the process parameters from the MOSIS 3 μm bulk CMOS process. Table 4-3 shows three sets of performance specifications, and SPICE simulation results for the circuit synthesized by OASYS for each. Figs. 4-2, 4-3, and 4-4 show the resulting sized circuit schematics. Circuit A requires minimal performance, and OASYS chose a simple 1-stage op amp topology. Circuit B stresses bandwidth, requiring a unity gain frequency of 10 MHz, and OASYS chose a simple 2-stage op amp topology. Circuit C stresses gain, requiring 90 dB of gain, and OASYS chose a 2-stage op amp topology but cascoded several of the sub-blocks. In all cases, simulation verifies that the op amp performance is quite close to the requested performance. Notice that a change in an input specification can result not only in different device sizing, but in the selection of an entirely different circuit topology, For all these test cases, the CPU time is very modest, usually less than 5 CPU seconds per circuit.

| Parameter | Units | Circuit A Style: OTA | | Circuit B Style: Simple 2-Stage | | Circuit C Style: Cascoded 2-Stage | |
|---|---|---|---|---|---|---|---|
| | | Spec | SPICE Result | Spec | SPICE Result | Spec | SPICE Result |
| Gain | dB | ≥ 60 | 62 | ≥ 60 | 76 | ≥ 90 | 96 |
| Unity Gain Freq. | MHz | ≥ 2 | 2.03 | ≥ 10 | 11.2 | ≥ 10 | 12.6 |
| Phase Margin | Degrees | ≥ 45 | 83 | ≥ 45 | 46 | ≥ 50 | 47 |
| Slew Rate | V/µs | ≥ 2.0 | +2.08/-2.10 | ≥ 2.0 | +13.4/-14.6 | ≥ 2.0 | +22.31/-19.6 |
| Load Capacitance | pF | 5 | - | 5 | - | 5 | - |
| Supply Voltage | V | ±2.5 | - | ±2.5 | - | ±2.5 | - |
| CMRR | dB | - | 93 | - | 83 | - | 119 |
| Power | mW | - | .193 | - | 1.12 | - | 1.53 |
| Estimated Area | $\mu m^2$ | - | 3020 | - | 6200 | - | 16200 |
| Input Devices (selected by OASYS) | - | - | PMOS | - | PMOS | - | NMOS |

**Table 4-3:** Comparison of 3 Different Designs.

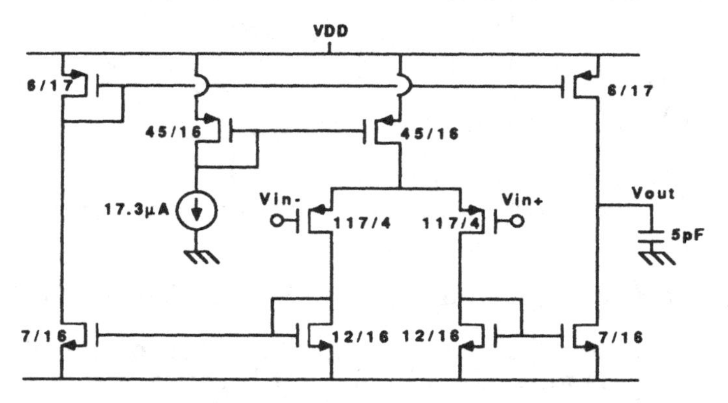


**Figure 4-2:** 1-Stage Op Amp Synthesized By OASYS.

# 5. DESIGN SPACE EXPLORATION

To picture the design space of an analog circuit, imagine an N-dimensional hypercube where each performance specification is assigned to a separate axis. Each point in this space represents a complete set of performance specifications for the design. The volume of designs that is achievable using a given set of topologies and design strategies can be regarded as the

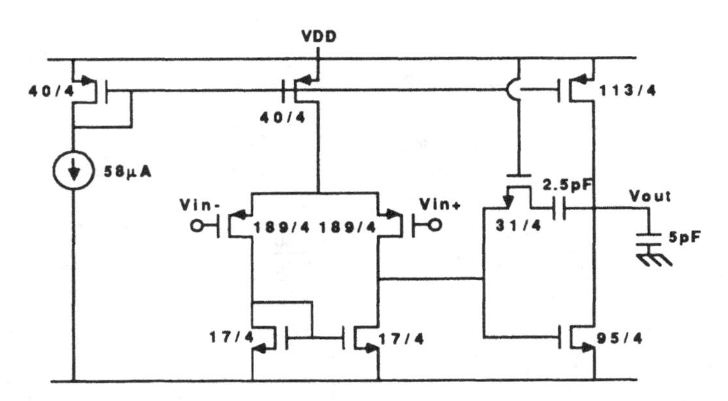


**Figure 4-3:** High Bandwidth Op Amp Synthesized By OASYS.

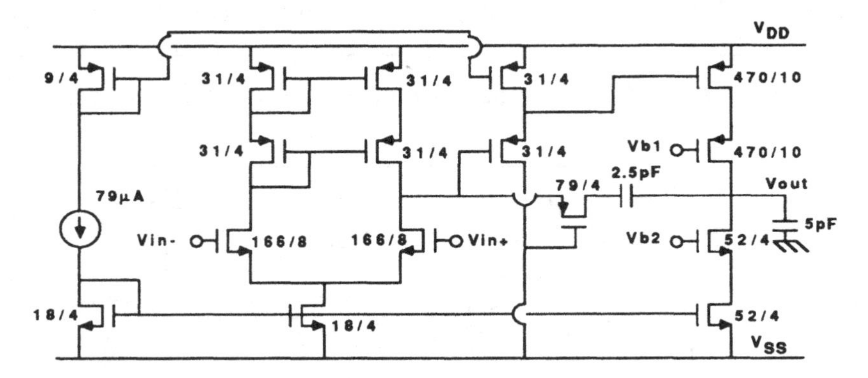


**Figure 4-4:** High Gain Op Amp Synthesized By OASYS.

designable space. Understanding the shape of the designable space for various modules is important because the system designer can use this information to quantify a tradeoff of performance requirements between modules. For example, if we place a tighter specification on the input offset voltage of an op amp in a dual slope A/D converter, then we can place a more relaxed specification on the input offset voltage of the comparator.

One approach to exploring the design space is to start by fixing some performance specifications, varying the remaining ones, and examining the area and power required for these designs. An example of this approach is shown in Figs 5-1 and 5-2. The two surfaces show the estimated power dissipation and estimated area over a range of bandwidth and slew rate specifications; all other parameters are fixed. Notice that as the specifications are changed, OASYS automatically changes its selection of design styles (e.g., 1-stage or 2-stage at the op amp level), in order to minimize power and area. Note, because of the hierarchical structure of the design style selection, not all 2-stage op amps have the same topology -- some will employ cascodes on various sub-blocks and some will not.

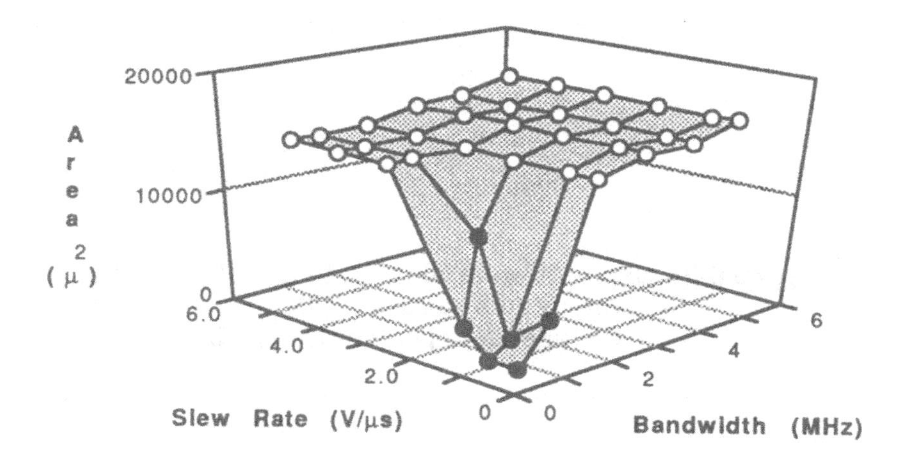


**Figure 5-1:** Exploration of Slew Rate and Bandwidth versus Active Area. Filled circles are 1-Stage and open circles are 2-Stage.

When one is not interested in area or power, but rather the possible tradeoff between two conflicting specifications, a second form of exploration, threshold hunting, is applicable. Fig. 5-3 shows, for the 2-stage design style, the region of feasible bandwidth specifications over a range of gain specifications; all other parameters are fixed. From this curve one sees the maximum possible tradeoff between gain and bandwidth -- at the set of values chosen for the rest of the performance specifications. This curve was determined by first increasing the gain at low bandwidth until OASYS could not perform the design (failed). From this point the bandwidth was increased slightly and the gain was started at its previous value and decreased until OASYS was able to complete the design successfully. This process is repeated until the tradeoff between gain and bandwidth has been determined over the desired range.

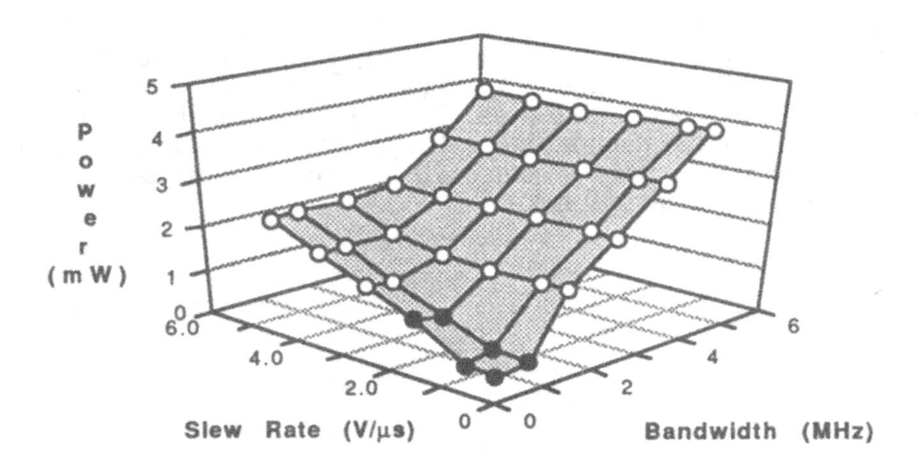


**Figure 5-2:** Exploration of Slew Rate and Bandwidth versus Active Area. Filled circles are 1-Stage and open circles are 2-Stage.

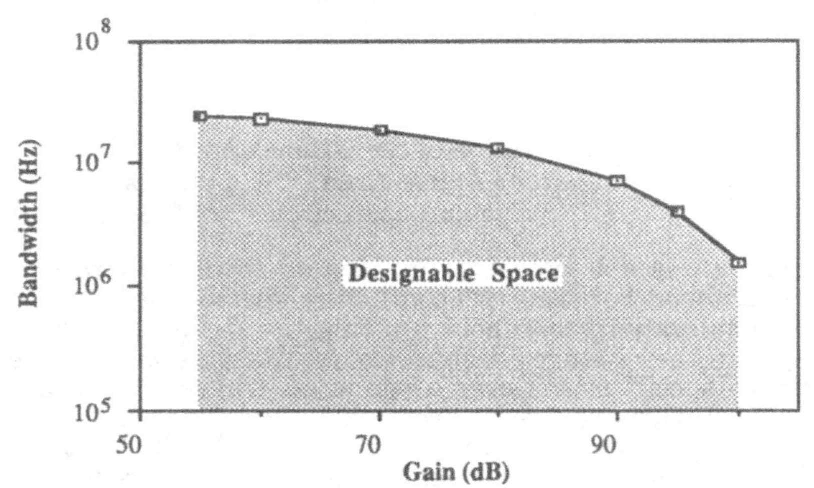


**Figure 5-3:** Threshold Hunting for the Edge of the Designable Space.

One final point, since OASYS works from process parameters as well as performance parameters, it is possible to explore the change in area and power with them as well. Fig. 5-4 shows the effects of changing threshold voltage on the area and power used by an op amp.

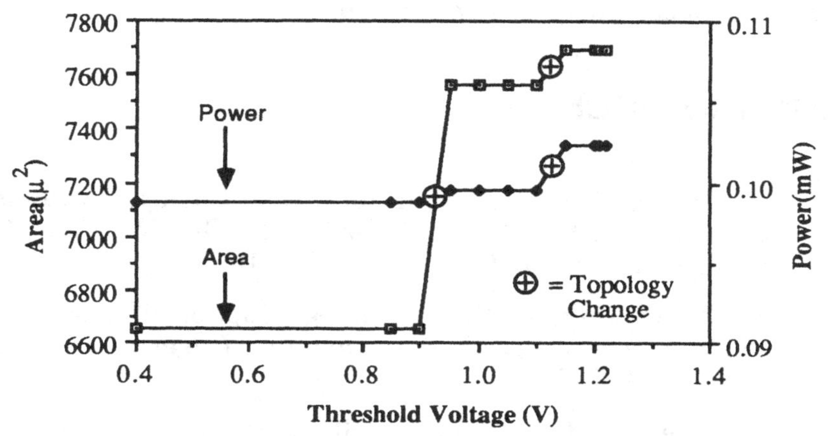


**Figure 5-4:** Exploration of Threshold Voltage Variations on Active Area and Power.

# 6. CONCLUSIONS

A general hierarchical framework for analog circuit synthesis has been developed and tested by incorporating domain knowledge specific to the design of CMOS op amps. The CAD tool, OASYS, takes performance specifications and a process technology file and generates sized transistor-level schematics. The performance of specific designs for many of these analog circuit topologies have been verified by measurements of fabricated devices [15] and others have been verified by circuit simulation. The extensibility of this method to analog circuits other than op amps has been verified by developing domain knowledge for the design of high speed regenerative comparators [28].

We see three important applications for a framework of this type. First, the ASIC system designer could incorporate these analog blocks without having to understand their design. Second, the analog circuit designer can this type of tool to carry out mundane portions of a design task leaving more time to concentrate on the challenging and new portions. Finally, because the automatic designs can be performed quickly, a large number of similar designs can be carried out in order to optimize tradeoffs between performance parameters. This process will be referred to as design space exploration.

## 7. REFERENCES


[1] D. Garrod, R.A. Rutenbar and L.R. Carley, "Automatic Layout of Custom Analog Cells in ANAGRAM", *Proc. 1988 IEEE Int'l Conf. on CAD*, November 1988.

[2] T. W. Pickerrell, "New Analog Capabilities on Semi-Custom CMOS", *Proc. Custom Integrated Circuit Conference*, 1983.

[3] T. Pletersek et al, "Analog LSI Design with CMOS Standard Cells", *Proc. IEEE Custom Integrated Circuit Conference*, 1985.

[4] G. I. Serhan, "Automated Design of Analog LSI", *Proc. IEEE Custom Integrated Circuit Conference*, 1985.

[5] R.L. Hedman, et al, "A Device-Level Auto Place and Wire Methodology for Analog and Digital Masterslices", *Digest Tech. Papers, IEEE Int'l Solid State Circuits Conf.*, Feb 1988.

[6] C.D. Kimble, A.E. Dunlop, G.F. Gross, V.L. Hein, M.Y. Luong, K.J. Stern, E.J. Swanson, "Autorouted Analog VLSI", *Proc. Custom Integrated Circuit Conference*, 1985.

[7] W.J. Helms and K.C. Russel, "Switched Capacitor Filter Compiler", *Proc. Custom Integrated Circuit Conference*, 1986.

[8] H. Yaghutiel, A. Sangiovanni-Vincentelli and P.R. Gray, "A Methodology for Automated Layout of Switched Capacitor Filters", *Proc. Int'l Conf. on CAD (ICCAD86)*, Nov. 1986.

[9] P. E. Allen and E. R. Macaluso, "AIDE2: An Automated Analog IC Design System", *Proc. IEEE Custom Integrated Circuiuts Conf.*, 1985.

[10] P. E. Allen and P. R. Barton, "A Silicon Compiler for Successive Approximation A/D and D/A Converters", *Proc. IEEE Custom Integrated Circuiuts Conf.*, 1986.

[11] J. Kuhn, "Analog Module Generators for Silicon Compilation", *VLSI Systems Design,* May, 1987.

[12] M. DeGrauwe, O. Nys, E. Vittoz, E. Dijkstra, J. Rijmenants, S. Cserveny and J. Sanchez, "An Analog Expert Design System", *Proceedings of the 1987 International Solid State Circuit Conference*, February 1987, pp. 212-213.

[13] M.G.R. DeGrauwe, et al, "IDAC: An Interactive Design Tool for Analog CMOS Circuits", *IEEE Jour. of Solid State Circuits*, Vol. SC-22, No. 6, December 1987.

[14] R. Harjani, R.A. Rutenbar and L.R. Carley, "A Prototype Framework for Knowledge-Based Analog Circuit Synthesis", *Proc. 24th ACM/IEEE Design Automation Conf.*, June 1987.

[15] R. Harjani, R.A. Rutenbar and L.R. Carley, "Analog Circuit Synthesis and Exploration in OASYS", *Proc. 1988 IEEE Int'l Conf. on Computer Design (ICCD)*, Oct. 1988.

[16] H.Y. Koh, C.H. Sequin and P.R. Gray, "Automatic Synthesis of Operational Amplifiers Based on Analytic Circuit Models", *Proc. 1987 IEEE Int'l Conf. on CAD,* Nov. 1987.

[17] H.Y. Koh, C.H. Sequin and P.R. Gray, "Automatic Layout Generation for CMOS Operational Amplifiers", *Proc. 1988 IEEE Int'l Conf. on CAD*, Nov. 1988.

[18] E. Berkcan, M. d'Abreu and W. Laughton, "Analog Compilation Based on Succesive Decompositions", *Proc. 1988 ACM/IEEE Design'Auto. Conf.*, June 1987.

[19] A.H. Fung, D.J. Chen, Y.N. Lai and B.J. Sheu, "Knowledge-based Analog Circuit Synthesis with Flexible Architecture", *Proc. 1988 IEEE Int'l Conf. on Computer Design (ICCD)*, Oct. 1988.

[20] A. P. Brokaw, "Design Proceedure for Bandgap Reference Generators", *Analog Devices Internal Document*, 1972.

[21] Centre Suisse d'Electronique et de Microtechnique, *User's Guide: IDAC, Interactive Design of Analog Systems, Version 2.2,* Neuchatel, Switzerland, 1987.

[22] B. Nye, A. Sangiovanni-Vincentelli, J. Spoto and A. Tits, "DELIGHT.SPICE: An Optimization-Based System for the Design of Integrated Circuits", *Proc. Custom Integrated Circuit Conference*, 1983.

[23] R. J. Bowman and D. J. Lane, "A Knowledge-Based System for Analog Integrated Circuit Design", *Proc. IEEE Internat. Conf. on Computer-Aided Design*, 1985.

[24] A. Ressler, *A Circuit Grammer for Operational Amplifier Design,* PhD dissertation, Artificial Intelligence Laboratory, Massachusetts Institute of Technology, 1984.

[25] M. G. R. Degrauwe and W. M. C. Sansen, "The Current Efficiency of MOS Transconductance Amplifiers", *IEEE Journal of Solid-State Circuits*, Vol. SC-19, No. 3, June 1984.

[26] L. W. Nagel, "SPICE2: A Computer Program to Simulate Semiconductor Circuits", ERL Memo ERL-M520, University of California, May, 1975.

[27] R. Harjani, R.A. Rutenbar and L.R. Carley, "OASYS: A Framework for Analog Circuit Synthesis", *IEEE Trans. CAD of ICs and Systems*, No. , 1989.

[28] R. Harjani, R.A. Rutenbar and L.R. Carley, "Analog Circuit Synthesis for Performance in OASYS", *Proc. 1988 IEEE Int'l Conf. on CAD*, November 1988.

# CHAPTER 4

# AUTOMATION OF ANALOG DESIGN PROCEDURES

Georges GIELEN and Willy SANSEN
Katholieke Universiteit Leuven-Elektrotechniek
94K Mercierlaan B-3030 Leuven
BELGIUM

## 1 INTRODUCTION

Advances in VLSI technology nowadays allow the realization of complex circuits and systems. ASIC's are moving towards the integration of complete systems on one chip, including both digital and analog parts. The use of computer-aided design tools has become indispensable to reduce the design time and cost. But whereas for digital VLSI a lot of design tools, methodologies and architectures have been developed, resulting now in complete digital silicon compilers, the first analog design tools and methodologies are just being introduced [1-7]. Most of the design is still carried out manually by analog experts, deriving simplified circuit expressions by hand and iteratively using numerical circuit simulators (such as SPICE) by trial and error. This explains why the analog part - though relatively small in area - takes more and more of the overall design time and cost of the present mixed analog-digital chips. This is especially true in high-performance applications requiring high frequencies, low noise and low distortion. Besides, due to the hand design and hand layout, errors frequently occur in the small analog section, leading to several reruns.

The design of high-performance analog circuits is known to be very knowledge-intensive. It strongly relies on the insight and expertise of the analog designer. This is mainly due to the complicated interaction between the device characteristics and the circuit performance. But also, insight into performance degrading effects such as noise and distortion is only obtained after long and tedious calculations and simulations. Hence, the analog CAD tools should be able to capture and formalize the existing design knowledge.

The *goals of computer-aided analog design tools* are therefore the following :

1. They have to shorten the overall design time, from the behavioral (sub)system description down to silicon.

2. They have to avoid errors in the design. The resulting silicon should be guaranteed first-time working (and testable!).

3. They must be able to predict performance of mixed analog-digital designs. Analog and digital blocks interact through the common substrate, common supply lines, capacitive couplings... Hence, power supply rejection ratio, thermal design... become very important.

4. They have to capture and formalize the existing knowledge of analog designers. Real analog experts are rare, hard to find and hard to keep. Their knowledge should be transferred to the CAD system.

5. They have to provide insight to the tool user. This allows the transfer of the built-in knowledge to new designers and allows to shorten their education time.

In order to realize these goals, all analog CAD tools are to be integrated into *one framework*, compatible with digital tools. Of great importance is the system's openness : expert designers should always be able to put in new knowledge and to ask explanation for the system's decisions. Hence analog design will always - even in an automated environment - be part of a learning process, in which the designer himself takes the ultimate decisions. Inexperienced (system) designers on the other hand can run the system in automatic mode, relying on the built-in knowledge.

Another important feature is *hierarchy*. In order to make system design feasible, the design methodology should be structured on several levels of hierarchy. For analog CAD however, these levels are not strictly defined. A possible organization could be :

- component level : transistor, capacitor...
- circuit level : operational amplifier, comparator...
- functional level : low pass filter, A/D converter...
- system level : modem chip, data-acquisition chip...

Analysis and synthesis tools should be present at all levels.

In the next Section, a systematic analog design methodology is presented to automate the design of functional analog blocks. It includes symbolic analysis of analog circuits, circuit sizing and optimization and analog layout generation. These tools are described in Sections 3, 4 and 5 respectively. Section 6 then presents concluding remarks.

# 2 SYSTEMATIC ANALOG DESIGN METHODOLOGY

Applications for telecom, audio, automotive, biomedicines . . . require the design of analog and mixed analog-digital systems. *At the system level*, this implies a partitioning of the system in (relatively non- interacting) building blocks. An example of a data-acquisition chip is shown in Fig. 1. The system description and specifications are translated down into specifications for the individual building blocks. These blocks (such as opamps, buffers, biquads...) are then designed and laid out according to the analog design methodology proposed in this Section. Next, the block layouts are combined and interconnected and the whole system is functionally verified. The router has to deal with typical analog constraints, such as minimization of parasitic capacitive and resistive load on sensitive nodes, avoidance of critical couplings, distribution of separate analog and digital supply lines... However, no real and mature tools have been presented at the system level yet. A general mixed-level mixed analog-digital system simulator is needed for the functional verification of analog-digital systems. An efficient analog-digital router is needed. And expert tools to aid designers with the system synthesis are needed. Also, the testability of mixed analog-digital systems still poses large problems.

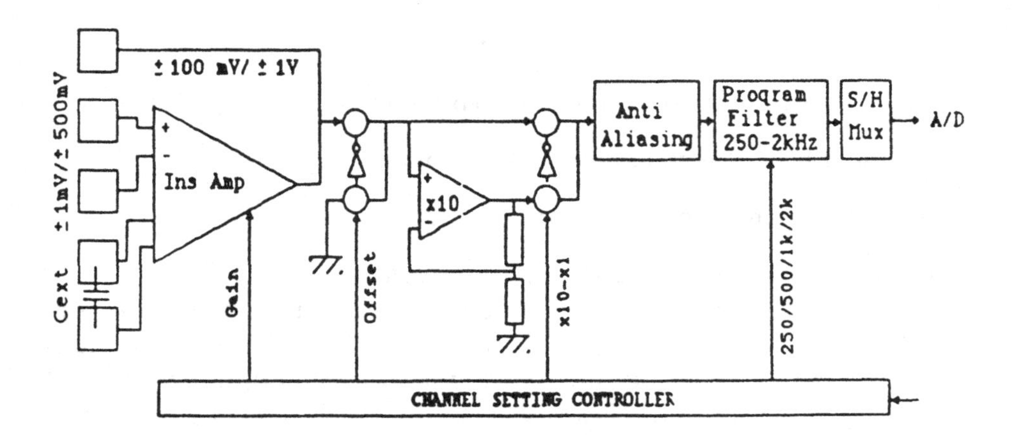


Figure 1: Data-acquisition chip for biomedical applications.

For the *automatic synthesis of analog building blocks*, several approaches and tools are published [2-7]. In this chapter, a new analog design methodology is presented [1]. It combines symbolic simulation and numerical optimization to allow an optimal circuit sizing and a fast inclusion of new topologies. The design methodology is schematically shown in Fig. 2. Generally, the sizing of an analog circuit from performance and technology specifications is formulated as an optimization problem, based on an analytic description of the circuit behavior.

The user first enters the performance specifications and selects a circuit topology and the technology process. In the future, an expert system is planned for the automatic selection of a topology depending on the specification values. The performance specifications are modified in the specification distortion routine to model various effects, such as temperature variations. The selected circuit topology is looked up in the database. If it is not present, an analytic circuit model is automatically generated by calls to the symbolic simulator ISAAC (Interactive Symbolic Analysis of Analog Circuits) [8-10]. The new model is then stored into the database.

Next, the OPTIMAN program (OPTIMization of ANalog circuits) [1] sizes all circuit elements based on the analytic model and the technology data, to satisfy the performance specifications. The degrees of

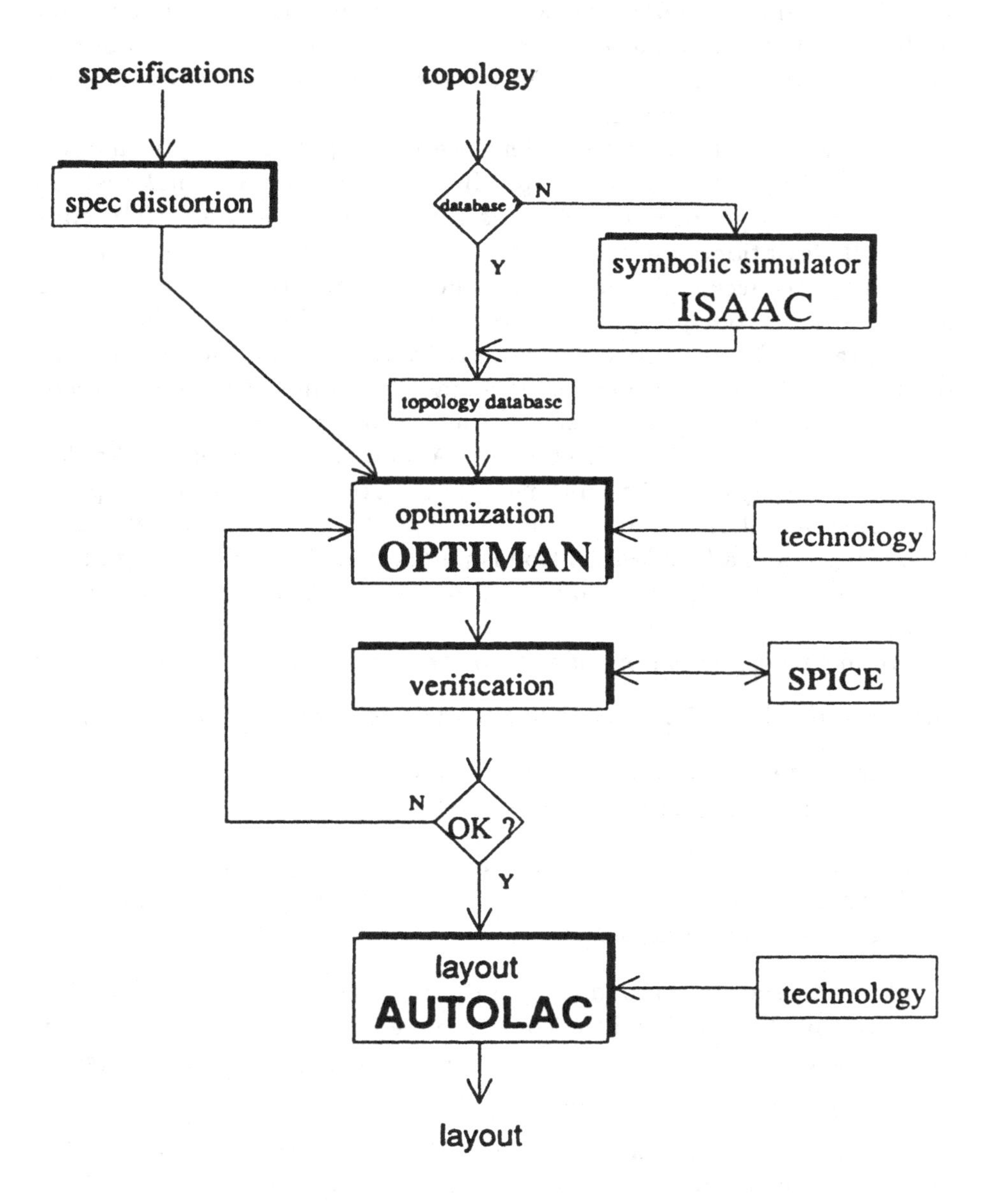


Figure 2: Design methodology for analog building blocks.

freedom in the design are used to minimize a user supplied cost function. For example, biomedical applications (such as data acquisition, intelligent sensors...) require minimal power consumption and minimal noise. Output buffers usually require large swing, low quiescent power, fast settling time and low distortion.

The design is then verified. The use of simplified analytic models speeds up the optimization and produces good, near optimal results. However, a final performance check is carried out with the SPICE simulator. If deviations are noticed, the optimization routine is recalled. If the design is accepted, the circuit schematic and the device sizes are passed to the layout program. If no acceptable solution can be found, the program indicates which characteristics are responsible for the failure. In the future, an automatic backtrace is planned to make changes to the existing topology or to select a new topology.

The layout generator AUTOLAC (AUTOmatic Layout of Analog Circuits) then lays out the building block, taking into account requirements such as minimum area, coupling, matching... Parasitic elements are extracted and a final verification is performed. If the design is not acceptable after layout, the circuit is resized or eventually a new topology is required.

Summarizing, *the main features of the new design methodology* are :

- it is independent of any particular circuit topology or technology.
- it allows the automatic inclusion of new topologies, due to the link with the symbolic simulator ISAAC.
- it is general and robust : the program can optimize any circuit that is described by an analytic model in acceptable CPU-times, also for circuits of practical size.
- it allows to specify a broad range of performance characteristics. The optimization goal and the constraints are fully tunable by the user. This also allows to optimize and explore a circuit in all dimensions of the analog design space.

The three main modules, the symbolic simulator, the optimization routine and the layout generator are described in the following sections. Knowledge is explicitly used in the topology selection routine and the analytic model generation routine (which calls ISAAC). All other routines, such as the layout, exploit analog design knowledge implicitly.

# 3 ISAAC: SYMBOLIC SIMULATION OF ANALOG CIRCUITS

## 3.1 Definition of symbolic analysis

The symbolic simulator ISAAC (Interactive Symbolic Analysis of Analog Circuits) [8-10] analyzes linear or linearized (small-signal) circuits in the frequency domain and returns analytic expressions with the complex frequency and (all or part of) the circuit elements represented by symbols. Time-continuous circuits are analyzed in the s-domain (Laplace domain), time-discrete circuits in the z-domain. Generally stated, the program derives network functions as symbolic rational forms in the complex frequency variable x (s or z), as given by :

$$H(x) = \frac{N(x)}{D(x)} = \frac{\sum_i x^i.a_i(p_1,\ldots,p_m)}{\sum_i x^i.b_i(p_1,\ldots,p_m)} \tag{1}$$

where the $a_i$ and $b_i$ are symbolic polynomial functions of the circuit elements $p_j$, that are represented by a symbol instead of a numerical value.

For semiconductor circuits, the exact expressions are usually long and complicated. Therefore, a heuristic approximation algorithm has been built in. It simplifies the expressions based on the relative magnitudes of the elements and returns the dominant terms in the result.

Example Consider for example the multi-phase switched-capacitor

integrator, shown in Fig. 3a [11]. The corresponding clocking scheme is depicted in Fig. 3b. The simplified transfer function with 25% error, $C_2 >> C_1$ and $A_1 >> 1$ is given by :

$$\frac{VOUT}{VIN} = \frac{C1\,(A2 - Z)}{C2\,Z\,(1 - Z)} \tag{2}$$

The simplified transfer functions from the opamp offset voltages to the integrator output are :

$$\frac{VOUT}{VOS1} = \frac{-A2 + Z - Z^2}{Z\,(1 - Z)} \tag{3}$$

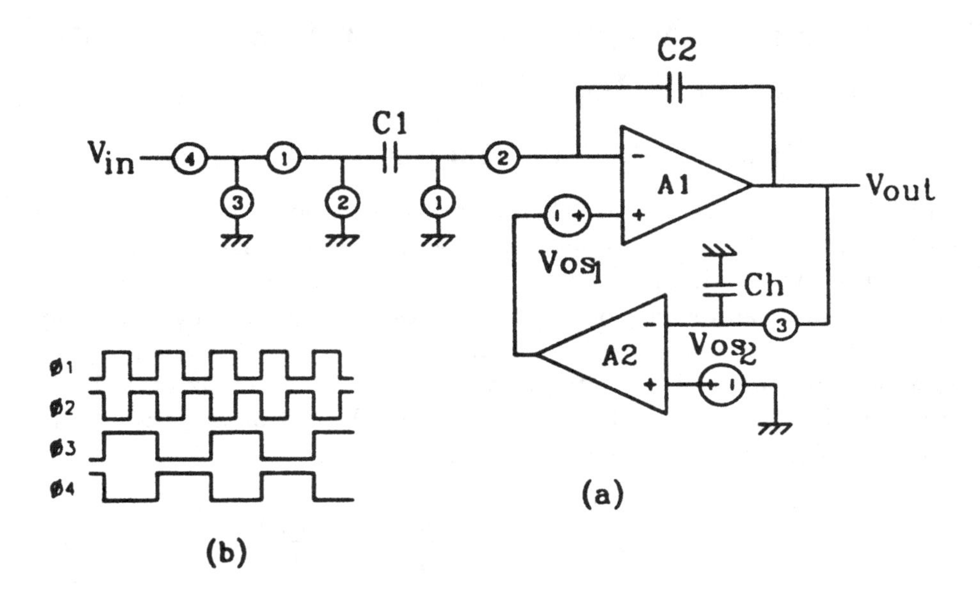


Figure 3: Switched-capacitor integrator with offset cancellation: (a) circuit schematic, (b) clocking scheme.

$$\frac{VOUT}{VOS2} = \frac{A2\,(-A2 + Z - Z^2)}{Z\,(1 - Z)} \tag{4}$$

The symbolic expressions (3) and (4) clearly show that the opamp offsets are canceled between the autozeroing period ($\Phi 3$ high) and the integration period ($\Phi 4$ high), on condition that $A_2 << 1$. These analyses took 2.8 sec of CPU time on a 1 Mips machine.

## 3.2 Symbolic versus numerical simulation

It is a tedious and time consuming process to obtain insight in a circuit by running a *numerical simulator* such as SPICE. Only numerical data are returned. The only conclusion that can be drawn from these numbers is whether the circuit meets the specified functional behavior or not, for the parameters provided in the input file. No indication is given which circuit elements essentially determine the observed performance. No potential problems are pointed out. No solutions are suggested when the circuit does not meet the specifications. In order to investigate the influence of parameter variations on the circuit performance, the parameters are to be varied for a repeated number of runs. Even then the sensitivity of the performance with respect to a given parameter is never fully understood. The situation is even worse for second-order phenomena, such as the power supply rejection ratio (PSRR), which strongly depends on component mismatching and layout overlap capacitances, which should be included in the input file. It can be concluded that numerical circuit simulators are mainly useful for functional (nominal) verification, once the circuit has already been designed.

On the other hand, to really gain insight into the circuit's behavior, the circuit has to be analyzed with the circuit elements represented by symbols. A *symbolic simulator* returns analytic expressions, which give the circuit performance (such as the gain of an opamp) in terms of element parameters (such as the transconductance of a transistor). These analytic expressions are valid whatever the element values are. The role of each parameter in the expression is clearly recognized. The analytic expressions also indicate the dominant design variables, especially when the expressions are simplified. They even show performance trade-offs and sensitivities to parameter variations. In this way, the designer learns about the performance of the circuit in a much more efficient way. The need for a symbolic simulator becomes even more stringent, when second-order characteristics such as the PSRR and the CMRR have to be obtained. It can be concluded that a symbolic simulator is a valuable analog design aid and forms an essential complement to numerical simulators.

Example

The differences between symbolic and numerical simulation are now illustrated in the following example [8]. Consider the bipolar cascode output stage, depicted in Fig. 4. A bootstrap capacitor $C_B$ has been added, which is supposed to increase the output impedance at higher frequencies. A SPICE result of the output impedance versus frequency for a $C_B$ value of 100 pF is shown in Fig. 5. The transistors are modeled by $r_\pi$ , $g_m$ and $g_o$ only. The plot confirms the expected impedance improvement. However, no indication is given by which elements the impedance levels at low and high frequencies or the two break points are determined. Even an experienced designer has difficulties to formally predict the influence of $C_B$ on the output impedance, although it is a

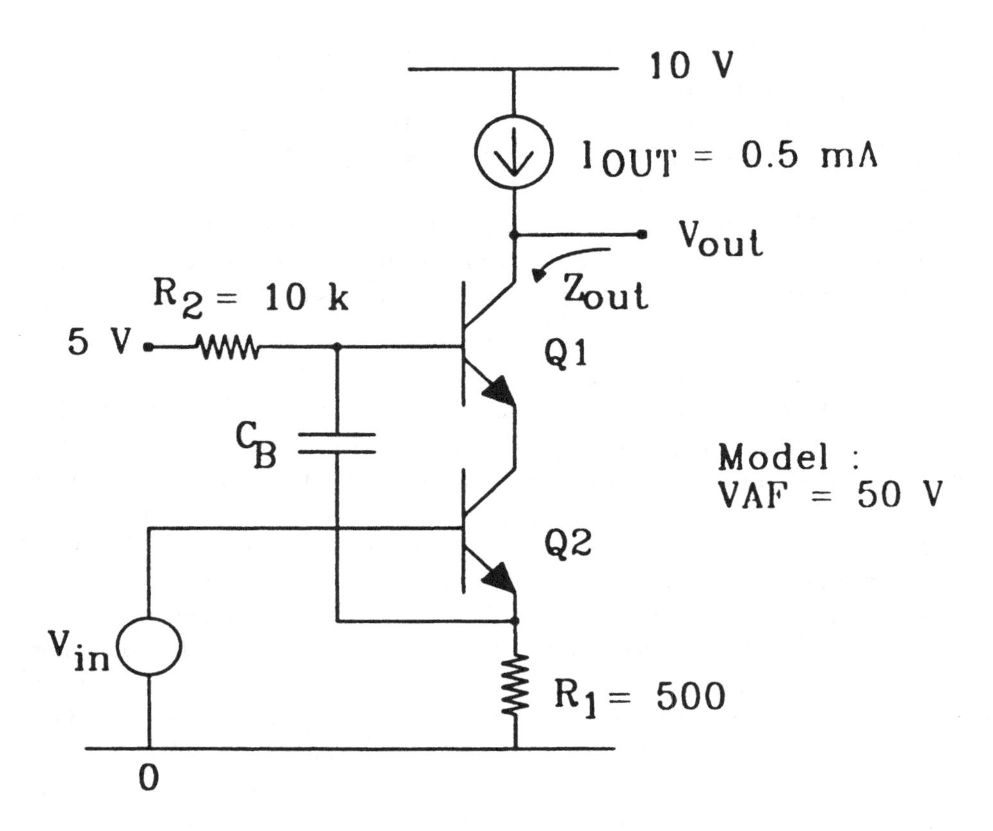


Figure 4: Bipolar cascode output stage with bootstrap capacitor.

simple two-transistor circuit.

The input file to the ISAAC program is shown below. It is stated in the well-known SPICE format (SWAP [12] for switched-capacitor circuits). In this way, compatibility with existing numerical simulators is established. The only difference is that numerical values do not have to be supplied.

```
BOOTCAS
*
VDD1 1 0 DC 10
VDD2 2 0 DC 5
```

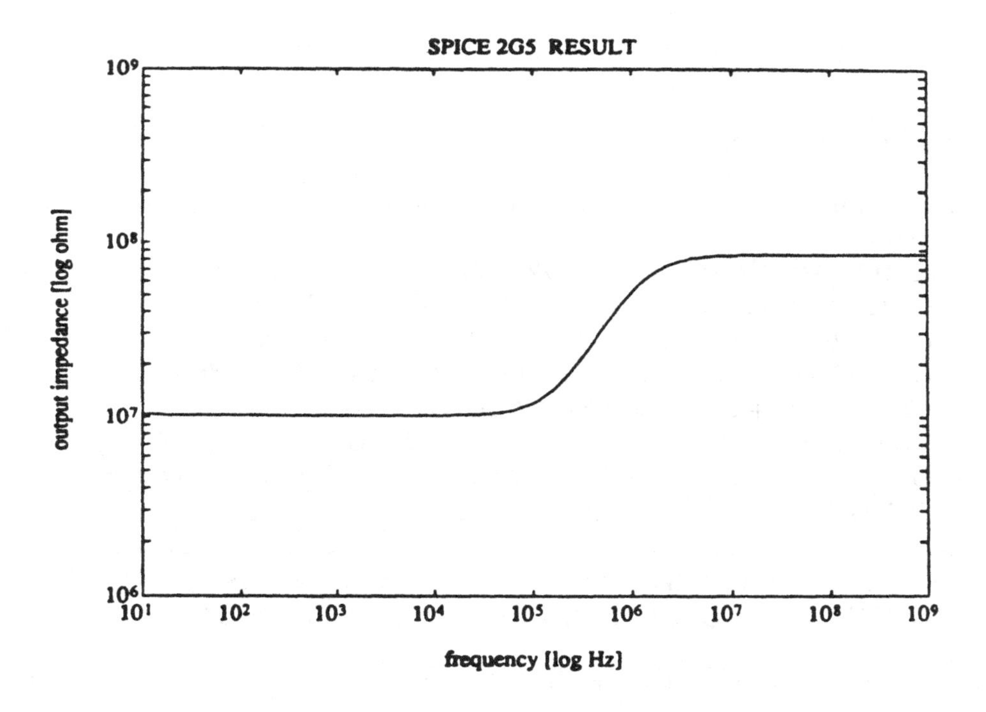


Figure 5: SPICE plot of the output impedance of the bipolar cascode stage.

```
Q1 4 3 5 0 NPN1
Q2 5 7 6 0 NPN2
R1 6 0 500
R2 2 3 10K
CB 3 6
VIN 7 0
IOUT 1 4
*
.MODEL NPN1 NPN VAF=50
```

The complete symbolic ISAAC expression for the output impedance is given by :

$$\frac{\begin{array}{l} G2GM1GM2 + G1G2GM1 + G2GM2G\pi1 + G2GM1G\pi2 \\ + GM2GO1G\pi1 + G1G2G\pi1 + G2GM2GO1 + G2GM1GO2 \\ + G1GO2G\pi1 + G2G\pi1G\pi2 + G1GO1G\pi1 + G1G2GO2 \\ + G1G2GO1 + GO2G\pi1G\pi2 + GO1G\pi1G\pi2 + G2GO2G\pi1 \\ + G2GO1G\pi2 + G2GO2G\pi2 + GO1GO2G\pi1 + G2GO1GO2 \\ + SCB(GM1GM2 + G1GM1 + GM1G\pi2 + G2GM1 + G1G\pi \\ + GM2GO1 + G\pi1G\pi2 + G1GO2 + G2G\pi1 + G1GO1 + GO1G\pi2 \\ + GO2G\pi2 + GO1G\pi1 + G2GO1 + G2GO2 + GO1GO2 \end{array}}{\begin{array}{l} GO1(G2GM2G\pi1 + G1G2G\pi1 + G2G\pi1G\pi2G1GO2G\pi1 + G1G2GO2 \\ + GO2G\pi1G\pi2 + G2GO2G\pi2 + G2GO2G\pi1 + SCB(G1G\pi1 \\ + G\pi1G\pi2 + G2G\pi1 + G1GO2 + GO2G\pi2 + G2GO2)) \end{array}} \tag{5}$$

This expression is already long for such a simple circuit. However, for a bipolar transistor it usually holds that $g_m >> g_\pi >> g_o$. Exploiting this information, the expressions can be simplified, thereby introducing some error. The maximum error percentage is supplied by the user. The simplified output impedance for 10% error is given by :

$$\frac{GM1\,(GM2\,+\,G1)\,(G2\,+\,SCB)}{GO1\,G\pi1\,(G2\,(GM2\,+\,G1)\,+\,SCB\,(G1\,+\,G\pi2))} \tag{6}$$

The approximation can be carried out further on, yielding simpler and simpler formulas, at the expense of more and more error. If a 25% error

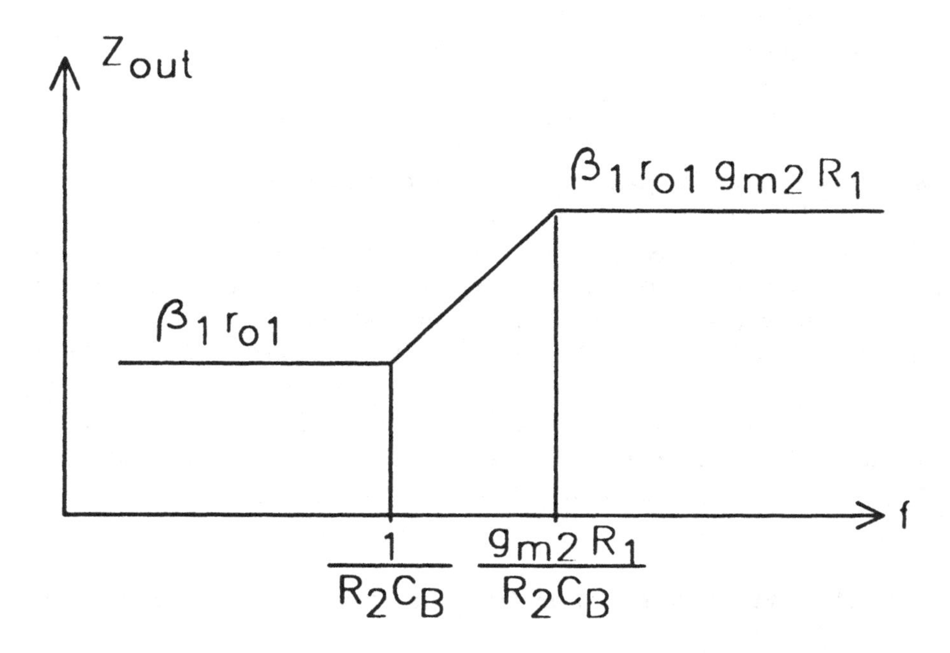


Figure 6: ISAAC plot of the output impedance of the bipolar cascode stage.

is allowed, the impedance is given by :

$$\frac{GM1\,GM2\,(G2 + S\,CB)}{GO1\,G\pi1\,(G2\,GM2 + S\,CB\,G1)} \tag{7}$$

At lower frequencies, this stage actually does not behave as a cascode stage, since the impedance at the emitter of Q1 is determined by $r_{\pi1}$ and not by $R_1.(g_{m2}\,r_{o2})$. However, at higher frequencies, the bootstrap capacitor $C_B$ connects $r_{\pi1}$ and $g_{o2}$ in parallel and the cascode formula holds again. The output impedance is now given by $R_1.(g_{m2}\,r_{\pi1}).(g_{m1}\,r_{o1})$. This behavior is summarized in Fig. 6. The symbolic simulator clearly gives more insight into the circuit than SPICE (Fig. 5) does.

## 3.3 Applications of symbolic analysis in analog design automation

### 3.3.1 Providing insight into circuit behavior

A symbolic simulator replaces the long and tedious hand calculations of the analog designer, especially for second-order effects, such as distortion and PSRR. The results are first-time correct and can be simplified to retain only the dominant terms. By examination of the resulting expressions, insight can be gained in the circuit behavior, as was illustrated in the previous example. This is useful for instruction and designer assistency, for students and novice designers, but even for experienced designers. The results are suitable for detection of the fundamental design parameters, for pole-zero extraction, for sensitivity and tolerance analysis [13] and even for fault diagnosis [14].

### 3.3.2 Analytic model generation for automatic circuit dimensioning

An important application of symbolic analysis is to integrate it as a CAD module into an analog silicon compiler [1,15]. Instead of using fixed cells, a more optimal solution towards analog module generation in terms of power, area and overall performance is the resizing or tuning of the circuit for every application. In our design strategy, this is done by characterizing all circuits by a full analytic model, which consists of performance equations and the independent design variables. These design variables are then optimized according to the specifications of the actual application [1].

However, as opposed to Degrauwe [2] who creates the formal circuit descriptions for IDAC after an in-depth analytic study, the circuit models are generated automatically by calling the symbolic simulator ISAAC. ISAAC then returns symbolic formulas for transfer functions, CMRR, PSRR, impedances, noise... In this way, it generates complete analytic models for building blocks such as amplifiers, filters, ladder networks... This allows the fast inclusion of new circuits and does not restrict the module generator to a fixed set of topologies. Also, the use of simplified formulas can strongly speed up the optimization for large circuits.

## 3.4 General description of the ISAAC program

ISAAC generates symbolic expressions for the AC characteristics of any analog integrated circuit (both time-continuous and switched- capacitor, CMOS, JFET and bipolar). The program is written in COMMON LISP and is implemented on VAX, Texas EXPLORER and Apollo DOMAIN computers. It provides the user with four analysis modes [10]. Time-continuous circuits, such as opamps and active RC filters, are analyzed in the s-domain. Time- discrete circuits, such as switched-capacitor filters, are analyzed in the z-domain. In addition, each phase of a time-discrete circuit can be analyzed separately in the s-domain with time constants or without time constants.

The *primitive elements* available in ISAAC are: independent voltage and current source, resistor, capacitor, inductor, current controlled current and voltage source, voltage controlled current and voltage source, nullator and norator, current and voltage meter, switch, short and ideal opamp. The user may define other non-primitive elements by means of a subcircuit statement. These subcircuits are then expanded into their primitive components during the analysis. Nonlinear devices such as transistors are also expanded into their small-signal model. These models are fully user controllable by the model statements. The default models for a bipolar and a CMOS transistor are shown in Fig.7.

As inputs, single and differential voltage and current sources can be selected. As outputs, single and differential node voltages and currents through elements can be asked.

Another feature is the *explicit representation of mismatch terms.* Consider for example transistors M1A and M1B in the CMOS Miller compensated two-stage opamp of Fig. 8. They are nominally identical. So, their small-signal components are represented by the same symbols. This accelerates the network function calculation and improves the readability of the result. However, for an accurate calculation of second-order effects such as the PSRR, device mismatching becomes important and explicit mismatch terms are added to the ISAAC analysis.

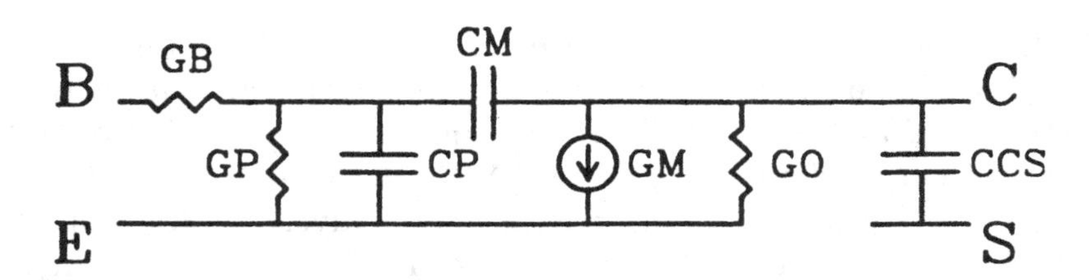


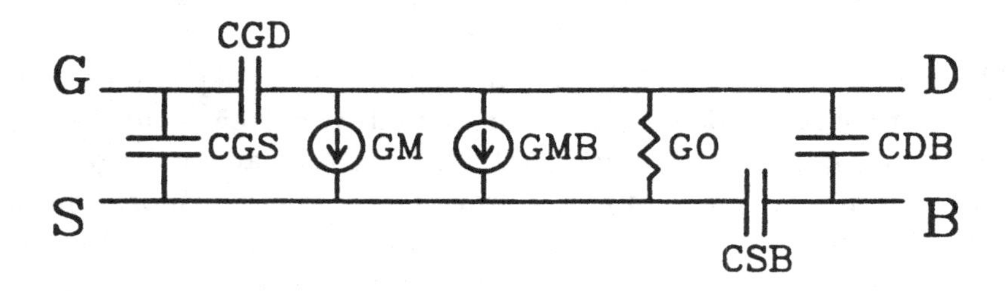


Figure 7: Default small-signal models for a bipolar and a CMOS transistor.

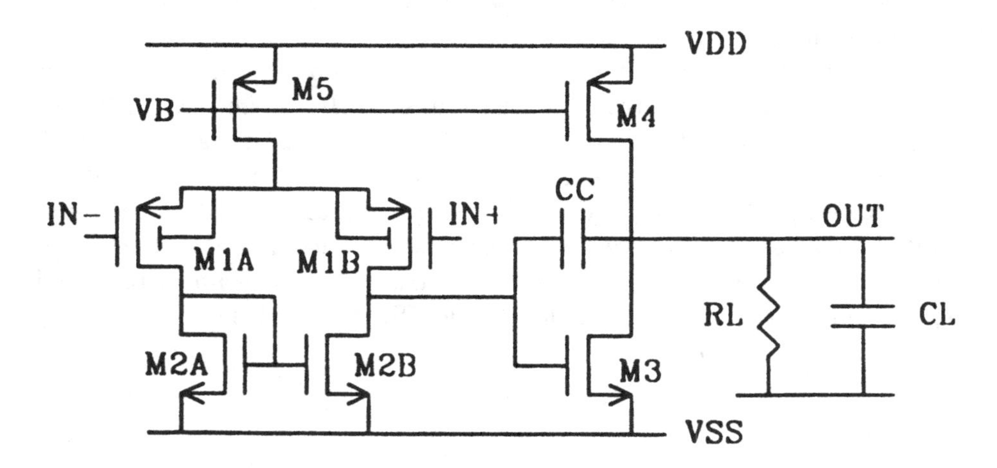


Figure 8: CMOS two-stage Miller compensated opamp.

Example

This is now illustrated in the PSRR expression for the positive supply voltage for the Miller opamp of Fig. 8. An approximation with 25% error yields :

$$\frac{\begin{array}{l}2GM1^2GM2GM3 + S(-2)CEQ - 2GM1GM1GM2 \\ + S^2(-1)CEQ - 2GM1(CEQ - 3GM1 + 2CGS1GM2) \\ + S^3(-1)CEQ - 2GM1(CEQ - 3CGS1 + CDB1CEQ - 3 \\ + 1/2CEQ - 1CEQ - 3 + CDB1CGS1) \\ + S^4CEQ - 2CGD1(CEQ - 3CGS1 + CDB1CEQ - 3 \\ + 1/2CEQ - 1CEQ - 3 + CDB1CGS1)\end{array}}{\begin{array}{l}2GM1GM2GO1GO4 + 2GM1GM2GO2GO4 + \\ \Delta GM2BAGM1GM3GO5 - \Delta GM1BAGM2GM3GO5 - \\ GM1GM3GO1GO5 - GM1GM3GO2GO5 + SGM1(2CEQ - \\ 2GM2GO4 + 2CEQ - 5GM2GO4 - CEQ - 3GM3GO5) + S^2(2CEQ - \\ 2CEQ - 4GM1GM2 + 2CEQ - 4CEQ - 5GM1GM2 - CEQ - \\ 1CEQ - 3GM1GM3 + 2CEQ - 2CEQ - 3GM1GO4 + CEQ - 2CEQ - \\ 3GM1GO5 + 2CEQ - 2CGS1GM2GO4) + S^3(2CEQ - 2CEQ - \\ 3CEQ - 4GM1 + CEQ - 1CEQ - 2CEQ - 3GM1 + 2CEQ - 2CEQ - \\ 4CGS1GM2 + 2CEQ - 3CEQ - 4CEQ - 5GM1) + S^4(2CEQ - 2CEQ - \\ 3CEQ - 4CGS1 + 2CDB1CEQ - 2CEQ - 3CEQ - 4 + 2CEQ - 3CEQ - \\ 4CEQ - 5CGS1 + CEQ - 1CEQ - 2CEQ - 3CEQ - 4 + CDB1CEQ - \\ 1CEQ - 2CEQ - 3 + 2CDB1CEQ - 2CEQ - 4CGS1)\end{array}}$$

with the following lumped elements being used :

$$\begin{aligned} CEQ - 1 &= CDB5 + CGD5 \\ CEQ - 2 &= CC + CGD3 \\ CEQ - 3 &= CDB2 + 2CGS2 \\ CEQ - 4 &= CDB4 + CGD4 \\ CEQ - 5 &= CDB2 + CGS3 \end{aligned} \tag{8}$$

At lower frequencies, a partial cancellation of two terms, $2\,g_{m1}\,g_{m2}\,g_{o4}\,(g_{o1} + g_{o2})$ and $-\,g_{m1}\,g_{m3}\,g_{o5}\,(g_{o1} + g_{o5})$, is noticed. These two terms correspond to the contribution of the first and the second stage. Also, the mismatch terms $\Delta g_{m2BA}\,g_{m1}\,g_{m3}\,g_{o5}$ and $-\Delta g_{m1BA}\,g_{m2}\,g_{m3}\,g_{o5}$ give a contribution, which is of the same order of magnitude as the other terms and which may not be neglected. At

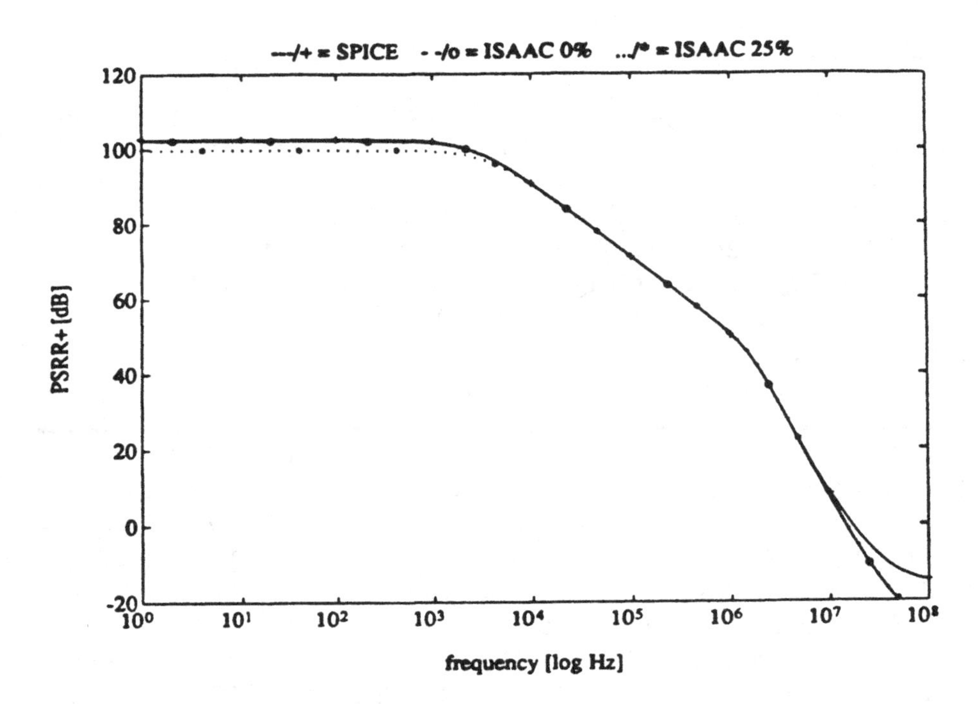


Figure 9: Comparison between SPICE and ISAAC for the $PSRR^+$ of the Miller opamp.

higher frequencies, the PSRR behavior is determined by the lumped capacitors CEQ-1, CEQ-2, CEQ-3, CEQ-4 and CEQ- 5. These are equivalent symbols for a series of capacitors in parallel.

It is also possible to *numerically evaluate the exact or simplified expression* over a user supplied frequency range. This yields plots which visualize the analytic behavior. These plots may be compared to SPICE output, to get an idea of the effective overall error introduced by the approximation. Fig. 9 compares the ISAAC expressions for 0% and 25% error with SPICE simulations of the PSRR+ for a typical design ($GBW = 1\,MHz, PM = 60°, C_L = 10\,pF$). In these analyses, mismatches of 1

Notice how close both ISAAC curves follow the SPICE result. Notice also that, although the maximum error percentage for expression (8) was 25%, the effective error for the magnitude of the $PSRR^+$ is much smaller over the whole frequency range. It can be concluded that most terms have only a minor contribution to the final result. The circuit behavior can be accurately described by simplified expressions and is determined by a small number of fundamental parameters only.

The analysis itself is based on the CMNA (Compacted Modified Nodal Analysis) formulation of the network equations. These equations are then solved by a dedicated and efficient algorithm, exploiting the sparse nature of the CMNA matrices. More *algorithmic details* can be found in [10]. More references on symbolic simulation can be found in [11]. Typical CPU times however on a 1 Mips machine are 8 and 26 seconds for the voltage gain of the Miller opamp (Fig. 8) and of a fifth-order switched-capacitor ladder filter respectively.

Conclusion

The symbolic simulator ISAAC is extensively used for the analysis of opamps, active filters and switched-capacitor circuits. It provides analog designers with more insight into the circuit behavior than numerical simulators do and has proven to be a valuable tool for instruction or designer assistency. Moreover, ISAAC is used for the generation of analytic circuit models, which are employed for automatic sizing in the non-fixed-topology analog module generator OPTIMAN. New functionality being planned in the future is : an approximate symbolic pole-zero extraction, sensitivity, nonlinear distortion, transient and large-signal analyses.

# 4 OPTIMAN: ANALOG DESIGN OPTIMIZATION

## 4.1 Analog circuit design based on analytic models

At the moment, more and more analog design strategies, frameworks and programs are being published [1-7]. In each of these programs, the analog circuits are resized for every application instead of using fixed cells. This guarantees an optimal solution in terms of area, power and overall performance, especially for ASIC's. The user first selects a particular circuit topology from the program's library. The program then sizes all circuit elements, to satisfy the performance constraints, possibly also optimizing an objective function.

The programs [2-4] all use some analytic description (design equations) of the circuit to perform the sizing and all exploit expert designer knowledge to some extent. However, the main drawback of these tools is that they are fixed-topology systems, limited to a fixed set of circuit topologies. None of them is able to automatically generate the appropriate analytic equations for a new topology. And the manual derivation of design equations is a tedious and error-prone job, especially for large circuits.

We present a *general approach towards automated analog circuit design*, which combines optimization and symbolic simulation [1]. To speed up the optimization, the circuits are characterized by an analytic model. However, whereas in OPASYN [4] the analytic model for a new topology must first be created by a good analog designer, this model is automatically generated in our approach by means of the symbolic simulator ISAAC [8-10]. This allows the automatic inclusion of new topologies into the system. Based on the analytic model, the OPTIMAN (OPTIMization of ANalog circuits) program then sizes the circuit topology to satisfy all performance constraints. The remaining degrees of freedom are used to optimize a general cost function. The optimization is based on simulated annealing, a statistical method to search the global optimum of a function [16]. OPASYN [4] on the other hand, uses a multi-start steepest gradient descent algorithm.

A straightforward approach to analog circuit design optimization is the combination of algorithmic optimization and circuit analysis techniques (simulation). This approach is applicable to a broad range of analog circuits and produces near optimal results. However, it is inefficient and costly in CPU time, due to the full simulation at every iteration.

The alternative, adopted in our program, is to replace the simulation by an *evaluation of analytic design equations*, which model the circuit behavior. This is based on the observation from the previous Section (see also Fig. 9) that most circuit characteristics, such as the gain or the phase margin of an opamp, are influenced by a small number of parameters only. All other circuit parameters have only a negligible influence. Moreover, the design equations are automatically derived by a circuit modeling routine, which calls the symbolic simulator ISAAC [8-10] to generate the AC characteristics for the circuit. The user may then trade off CPU-time consumption during optimization against accuracy, by introducing less or more error in the approximation.

## 4.2 The analog design formulation in OPTIMAN

The OPTIMAN program is based on a generalized formulation of the analog design problem. After the user has selected a circuit topology, the analytic model is generated, the independent design variables are automatically extracted and the program sizes all elements to satisfy the performance constraints, thereby optimizing a user defined design objective.

The optimization is based on *circuit models*, which analyticly model the circuit behavior. These circuit models contain a description of the circuit topology, a summary of the independent design variables, relationships between dependent and independent design variables, design equations relating the circuit performance to the design variables, additional analog expert knowledge about the circuit and other design constraints.

The *performance specifications* are treated in two different ways :

- performance objectives are incorporated in the goal function. They are minimized (or maximized) and the user has to supply a weighting coefficient and the maximum (or mimimum) value allowed. For example, a mixed power-area minimization may be performed with maximum values of $10\,mW$ and $10\,mm^2$.

- performance constraints are treated as inequality conditions during the optimization. For example, a minimum gain-bandwidth of 1 MHz and a gain of 60 dB may be required.

In addition, other boundary conditions can be added, such as the requirement on all capacitor values to be positive.

As *design variables*, all node voltages, currents and element values in the circuit are taken. These variables are reduced to an independent set, by means of general constraints (such as Kirchoff's laws), circuit specific constraints (such as matching information) and designer constraints (such as offset reduction rules). This independent set forms the fundamental design parameters, on which all characteristics of the circuit depend. For example, most of the characteristics of an operational amplifier depend on the bias current. Therefore this current can be used as an independent variable to optimize for example the gain-bandwidth. This is illustrated in Fig. 10 for a CMOS OTA. For maximum gain-bandwidth, a specific aspect ratio of the input transistors results. This maximum of course still depends on the current.

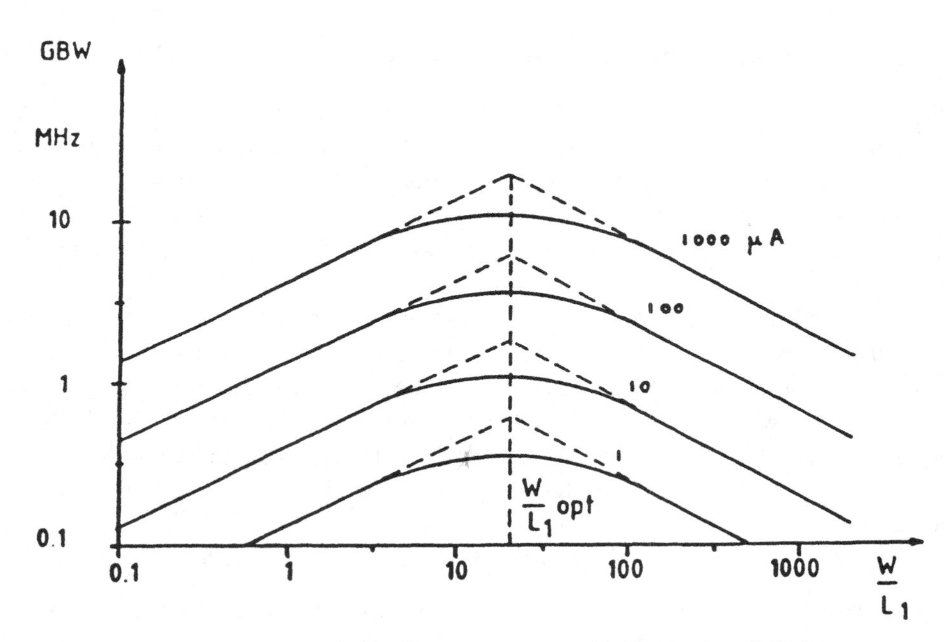


Figure 10: Gain-bandwidth optimization of a CMOS OTA

For the design of operational amplifiers, a useful set of design variables are those variables which determine the DC operating point of the circuit: bias currents, transistor saturation voltages, bias voltages (and transistor lengths). For example, the independent design variables for the CMOS folded-cascode OTA of Fig. 11 are : $I_1$, $(V_{GS}-V_T)_1$, $(V_{GS}-V_T)_3$, $(V_{GS}-V_T)_5$, $(V_{GS}-V_T)_7$, $(V_{GS}-V_T)_8$, $(V_{GS}-V_T)_{10}$, $V_{B1}$ and $V_{B2}$. From this DC operating point, all transistor dimensions W and L can be calculated, using a general transistor model. This transistor model may be a first- order model or a more sophisticated one. For example, the gain-bandwidth is determined by $g_{m1}$. However, the relationship between $g_{m1}$, $I_1$ and $(V_{GS}-V_T)_1$ is part of the transistor model.

The *optimization algorithm* itself is simulated annealing [16]. This is a general and robust method, based on random move generation and statistical move acceptance. Moves which lower the goal function are always accepted. Moves which increase the goal function are statistically accepted. The acceptance probability of up-hill moves is large in the beginning of the optimization, when the controlling temperature is high, but decreases with decreasing temperature. Since the routine allows uphill moves, it will find a solution close to the global optimum at the

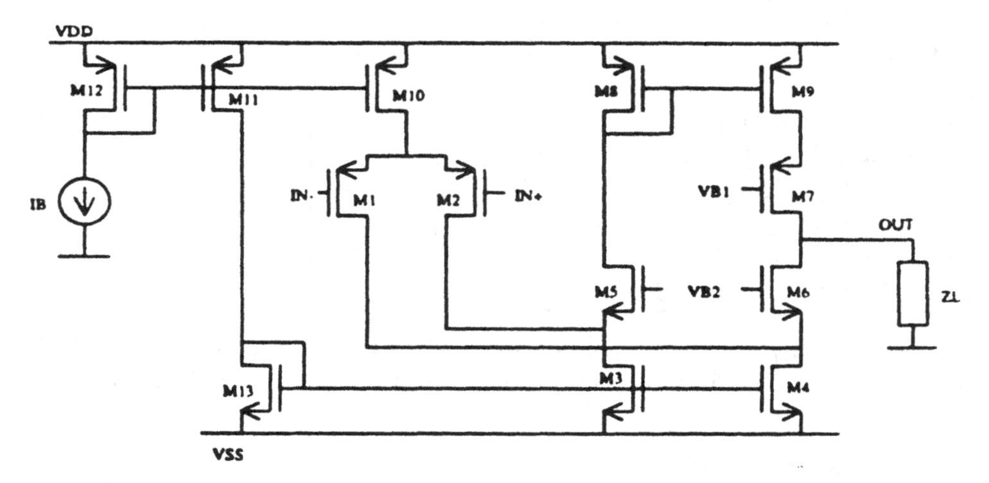


Figure 11: CMOS folded-cascode OTA.

expense of a large number of function evaluations. The simulated annealing implementation used is SAMURAI [17], an efficient kernel with fully adaptive temperature scaling and move range limitation and novel inner loop criterion. These features keep the CPU-time consumption still acceptable for large circuits.

The *general optimization* loop may then be summarized as follows : the program selects new values for the independent variables, calculates all dependent variables, checks if the corresponding design satisfies all boundary conditions and constraints, calculates the goal function and statistically accepts or rejects the new state. This loop is executed until convergence occurs at lower temperatures.

This may also be described in a more visual way. An analog circuit may be regarded as a multidimensional space, in which each performance specification is a separate dimension. The valid design space for a particular application consists of those points which satisfy the specification constraints. The OPTIMAN program searches in this valid design space for the point, which optimizes the goal function. For opamp designs, a possible goal function is a weighted combination of the 4 fundamental axes: power, area, dynamic range (limited by noise and distortion) and frequency range (gain-bandwidth).

| specification | spec | OPTIMAN | SPICE |
|---|---|---|---|
| GBW $[kHz]$ | 200 | 223 | 224 |
| power $[mW]$ | $< 1$ | 0.140 | 0.142 |
| gain $[dB]$ | $> 60$ | 83 | 97 |
| phase margin $[°]$ | $> 60$ | 89 | 86 |
| slew rate $[V/\mu s]$ | 0.13 | 0.28 | 0.28 |
| noise $[\mu VRMS]$ | $< 20$ | 16.1 | 17.6 |
| output range $[V]$ | 1 | 1.0 | 1.07 |
| input range $[V]$ | 1 | 1.0 | 1.09 |
| offset $[mV]$ | $< 0.5$ | 0.14 | 0.1 |

Table 1: Comparison of specifications, OPTIMAN results and SPICE simulations for a minimal power design of the folded-cascode OTA.

### 4.3 Design examples

The OPTIMAN program is written in VAX PASCAL and runs under VMS on a VAX750. The present topology database contains models for several frequently used CMOS opamps. The technology used is $3\mu m$ CMOS n-well.

Consider for example the CMOS folded-cascode OTA of Fig. 11. The design specifications are compared to OPTIMAN results and SPICE simulations in Table 1. Notice the close correspondence between all figures. Table 2 compares the final optimization variables for power minimization, noise minimization and gain-bandwidth maximization of the folded-cascode OTA. All these designs require less than 1 minute on a VAX750.

It can be concluded that OPTIMAN has successfully been applied to the design of CMOS opamps and switched-capacitor circuits. The examples demonstrate the usefulness and reliability of the approach. Inexperienced opamp designers can easily use the program, fully in batch mode. Expert circuit designers can easily insert new topologies.

## 5 ANALOG LAYOUT GENERATION WITH AUTOLAC

Once the circuit topology has been selected and the optimal device dimensions determined, the circuit still has to be laid out. This process is

| variable | power | noise | GBW |
|---|---|---|---|
| $I_1$ | 13 $\mu A$ | 323 $\mu A$ | 500 $\mu A$ |
| $(VGS - VT)_1$ | 0.20 V | 0.20 V | 0.20 V |
| $(VGS - VT)_3$ | 0.36 V | 0.50 V | 0.43 V |
| $(VGS - VT)_5$ | 0.20 V | 0.27 V | 0.27 V |
| $(VGS - VT)_7$ | 0.27 V | 0.20 V | 0.35 V |
| $(VGS - VT)_8$ | 0.34 V | 0.50 V | 0.33 V |
| $V_{B1}$ | 0.3 V | 0.2 V | 0.1 V |
| $V_{B2}$ | -0.1 V | -0.2 V | -0.6 V |

Table 2: Optimization variables for power minimization, noise minimization and GBW maximization of the folded-cascode OTA.

split up into four different steps:

- selection of structural entities
- optimal placement and generation of the entities
- analog interconnection of the entities (by an analog router)
- final compaction

In this way, a near optimal layout can be generated for every application, taking into account typical analog constraints. A program, implementing these different steps, is currently under development. It is called AUTOLAC (AUTOmatic Layout of Analog Circuits). It is programmed on VAX 750 in VAX Pascal and runs in front of the symbolic layout program CAMELEON [18].

The *structural entities* (also called cells) are functional groups of elements: resistors, capacitors, single transistors, two or more matching transistors... Matching elements are taken together in order not to overconstrain the placement routine. All cells can be laid out in several ways. For example, four different layouts for a single transistor are shown in Fig. 12.

For all cells and all layout forms, parametric and technology independent expressions have been generated for the cell height and width as a function of the R, C or W and L value. Before the placement optimization, all possible layouts for a given cell are combined into the bounding curve of this cell, as shown in Fig. 13.

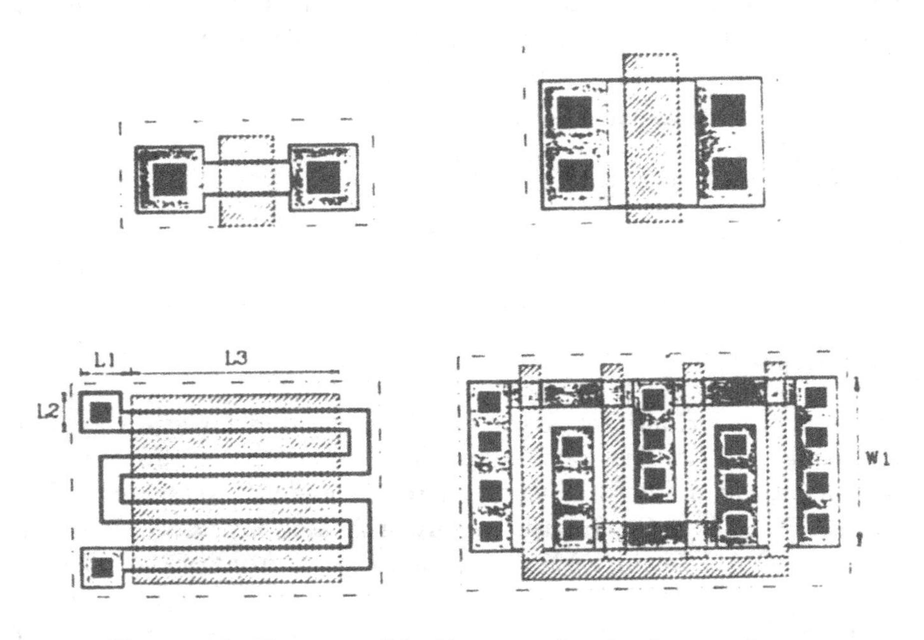


Figure 12: Four possible layouts of a single transistor .

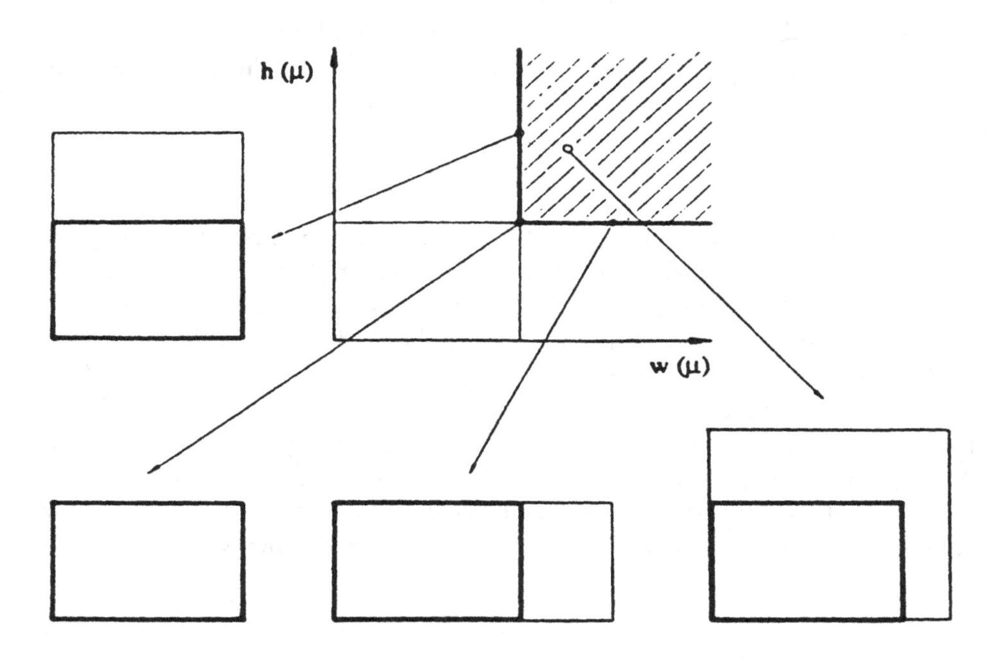


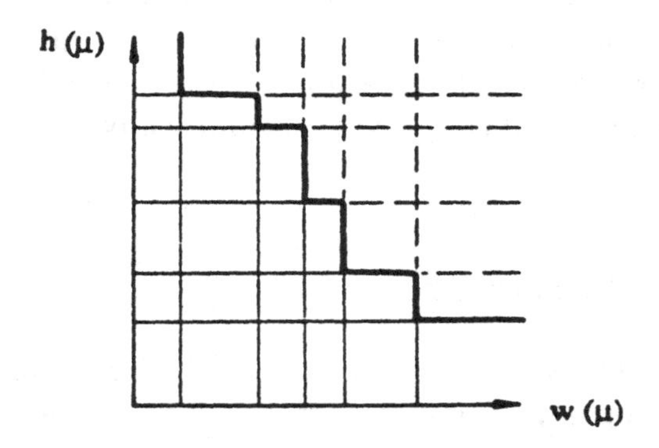


Figure 13: Bounding curve of a cell with several possible layout forms.

The *placement* itself is based on a slicing structure, which characterizes the relative placement of all cells in the global layout. This slicing structure is then reorganized iteratively by the optimization routine, in order to find the structure which minimizes the goal function. Again, the optimization algorithm is simulated annealing [16]. The goal function consists of an area and an interconnection term :

$$F = A + \lambda W \tag{9}$$

For the placement, a slicing structure is used since this implies only a relative (not an absolute) placement, which strongly reduces the search space. Also, the method allows cells with flexible dimensions. This is the case, since the structural entities can be laid out in different ways, with a different width and height.

The bounding curve of the global layout is obtained by combining the bounding curves of the individual cells, according to the slicing tree. At the end of the optimization, the point with minimum area on the final bounding curve is selected and the actual widths and heights for each cell can be derived by tracing the slicing tree downwards.

Example

Consider for example the CMOS folded-cascode opamp of Fig. 14.

It is divided in several functional entities (current mirrors, differential pair...). A compact placement result is shown in Fig. 15.

The analog router however is still under development. Hence, the chip shown in Fig. 16 - which corresponds to the optimization example of the previous Section (Table 1) - was automatically placed but manually routed.

# 6 CONCLUSIONS

A design methodology is presented for the design of functional analog building blocks (opamps, comparators... ). The method combines symbolic simulation and numerical optimization to allow an optimal circuit sizing and a fast inclusion of new topologies. The circuit optimization is based on analytic models, which characterize the circuit performance and which are generated by the symbolic simulator ISAAC.

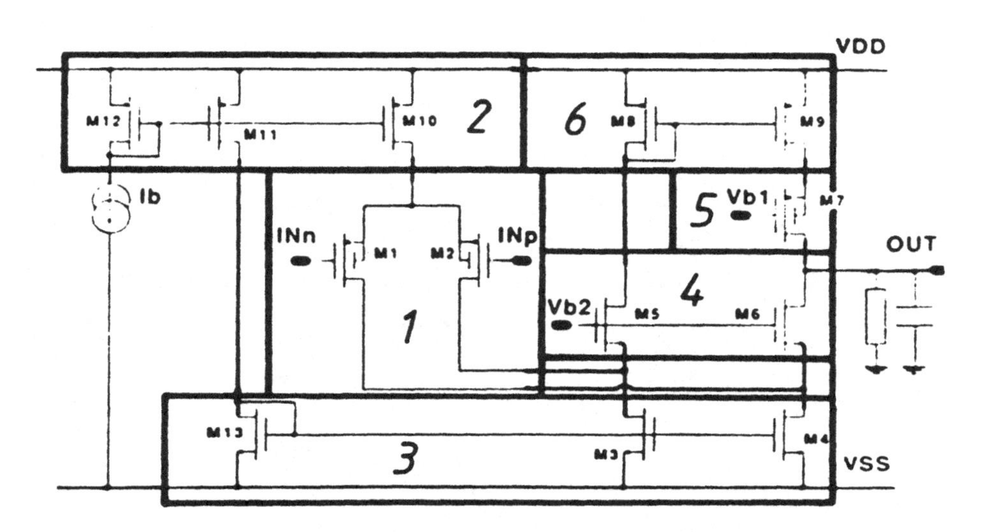


Figure 14: Division of a CMOS folded-cascode OTA in functional entities.

ISAAC generates symbolic expressions for all AC characteristics of both time-continuous and time-discrete circuits. It returns both exact and simplified analytic expressions. In this way, it provides analog designers with more insight into the circuit behavior than numerical simulators do. ISAAC is also used for the generation of analytic circuit models, which are employed for automatic sizing in the optimization program OPTIMAN.

OPTIMAN is based on a general formulation of analog design as an optimization problem. The program determines all independent circuit parameters in order to optimize a user defined objective, thereby satisfying all performance constraints. The optimization algorithm is simulated annealing.

The design is then passed to the layout generator AUTOLAC. First, all functional groups of elements in the circuit schematic are recognized. These groups are then optimally placed, by a routine based on slicing structures and bounding curves. Finally, the functional element groups are interconnected. The analog router however is still under development.

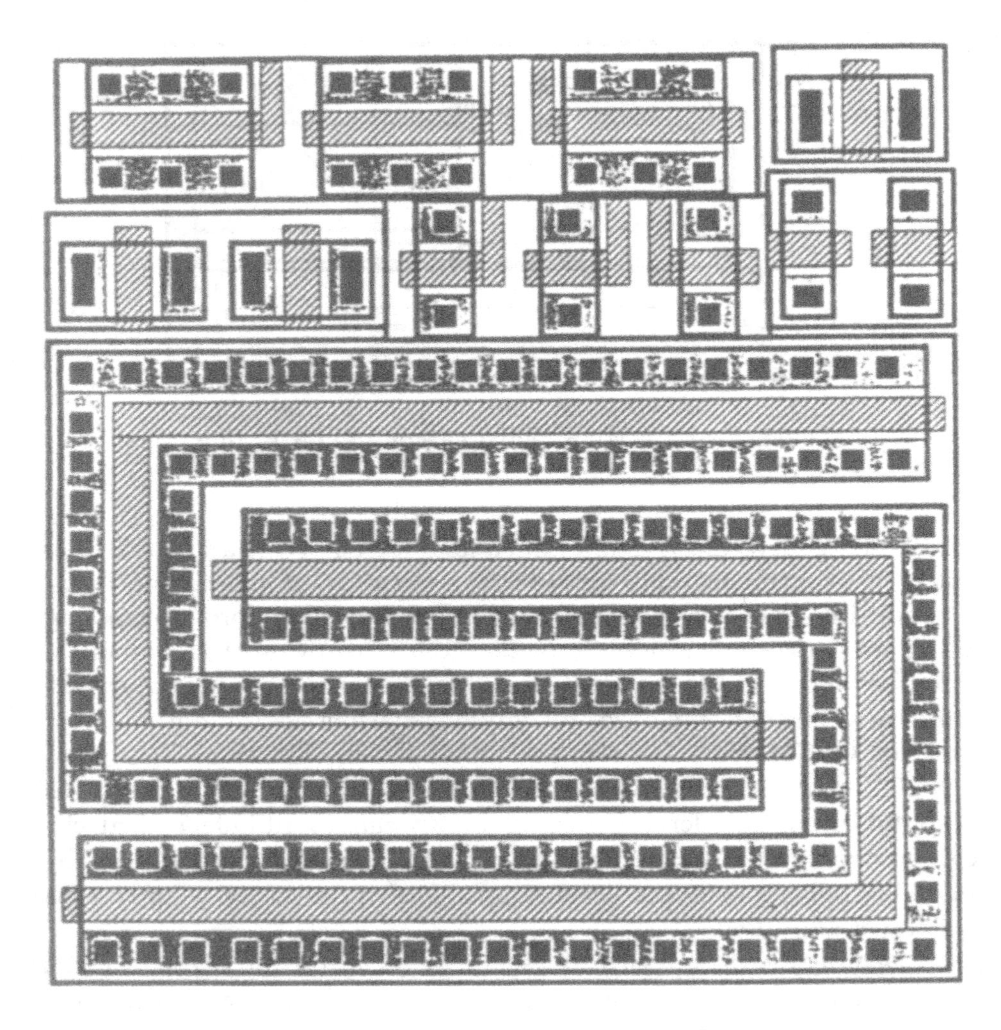

Figure 15: Placement result for a folded-cascode OTA.

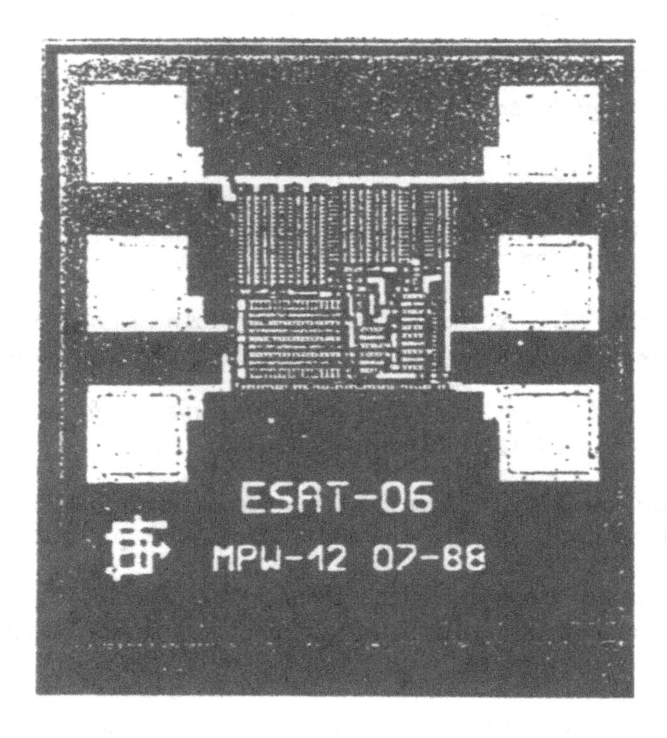


Figure 16: Chip photograph of a folded-cascode OTA.

## References

[1] Georges Gielen, Herman Walscharts, Willy Sansen : *"Analog design automation based on analytic simulation"* - proceedings ESSCIRC 1989.

[2] M. Degrauwe et al. : *"IDAC: an interactive design tool for analog integrated circuits"* - IEEE Journal of Solid-State Circuits, Vol. SC-22, No. 6, December 1987, pp. 1106-1116.

[3] R. Harjani, R. A. Rutenbar, L. R. Carley : *"A prototype framework for knowledge-based analog circuit synthesis"* - proc. DAC 1987, pp. 42-49.

[4] H. Y. Koh, C. H. Sequin, P. R. Gray : *"Automatic synthesis of operational amplifiers based on analytic circuit models"* - proc. ICCAD 1987, pp. 502-505.

[5] E. Berkcan, M. d'Abreu, W. Laughton : *"Analog compilation based on successive decompositions"* - proc. DAC 1988, pp. 369-375.

[6] F. M. El-Turky : *"BLADES: an expert system for analog circuit design"* - proc. ISCAS 1986, pp. 552-555.

[7] R. J. Bowman, D. J. Lane : *"A knowledge-based system for analog integrated circuit design"* - proc. ICCAD 1986, pp. 210-212.

[8] Willy Sansen, Georges Gielen, Herman Walscharts : *"Symbolic simulator for analog circuits"* - proc. ISSCC 1989, pp. 204-205.

[9] Herman Walscharts, Georges Gielen, Willy Sansen: *"Symbolic simulation of analog circuits in s- and z-domain"* - proc. ISCAS 1989, pp. 814-817.

[10] Georges Gielen, Herman Walscharts, Willy Sansen : *"ISAAC: a symbolic simulator for analog integrated circuits"* - submitted for publication to Journal of Solid-State Circuits.

[11] J. Robert et al. : *"Offset and clock-feedthrough compensated SC integrators"* - proceedings ISCAS, pp. 817-818, 1986.

[12] *"SWAP 2.2 Reference Manual"* - Silvar Lisco, Heverlee, Belgium, Doc. M-037-2, 1983.

[13] Kishore Singhal, Jiri Vlach : *"Symbolic analysis of analog and digital circuits"* - IEEE Transactions on Circuits and Systems, Vol. CAS-24, No. 11, pp. 598-609, November 1977.

[14] A. Liberatore, S. Manetti : *"SAPEC - a personal computer program for the symbolic analysis of electric circuits"* - proceedings ISCAS, pp. 897-900, 1988.

[15] A. Konczykowska, M. Bon : *"Automated design software for switched-capacitor IC's with symbolic simulator SCYMBAL"* - proceedings 25th ACM/IEEE Design Automation Conference, pp. 363-368, 1988.

[16] S. Kirkpatrick, C. D. Gelatt Jr., M. P. Vecchi : *"Optimization by simulated annealing"* - Science, Vol. 220, No. 4598, 13 May 1983, pp. 671-680.

[17] Francky Catthoor, Joos Vandewalle, Hugo De Man : *"SAMURAI: a general and efficient simulated-annealing schedule with fully adaptive annealing parameters"* - Integration, the VLSI Journal (North Holland), No. 6, June 1988, pp. 147-178.

[18] Kris Croes, Hugo De Man, Paul Six : *"CAMELEON, a process tolerant symbolic layout system"* - proc. ESSCIRC 1987, pp. 193-196.

## Acknowledgements

The authors are grateful to Philips Industries, the Netherlands, for the logistic support. Special thanks are addressed to Herman Walscharts (co-author of ISAAC), Koen Swings (many useful ideas and discussions), Geert Van Hecke and Carl Verdonck (contributions to OPTIMAN), Koenraad Van Schuylenbergh and Rudi Verbeeck (contributions to AUTOLAC).

# CHAPTER 5

# ANALOG DESIGN TOOLS FOR HIGH-FREQUENCY APPLICATIONS

Dr. Sorin A. Huss
AEG Aktiengesellschaft
Design Center for Integrated Circuits
Sedanstr.lo D-7900 Ulm
FR GERMANY

This chapter focuses on automated design tools for high-speed and non-linear bipolar analog integrated circuits in a frequency range up to 3 GHz. These tools are part of an actual design system developed for industrial applications. Several design examples are included to highlight the main features of these CAE tools.

## 1 INTRODUCTION

The purpose of this chapter is to give an overview of an actual design environment for analog integrated circuits at an industrial site, which is not part of a semiconductor factory, but acts as a central design center for a major systems house.

The target of this design environment is mainly (but not restricted to) the engineering of bipolar analog integrated circuits operating in a frequency range of up to 3 GHz, whereas non-linear circuits are of special interest. These circuits are key components of e. g. high-speed communication systems or radar signal processing units.

In the area of digital semi-custom integrated circuits the design tools and the available libraries have meanwhile matured yielding a highly automised approach to functional circuits at the cost of more or less stringent restrictions of an individual design such as a limited use of transmission gate cells in gate array based CMOS applications. Characteristic of this approach to semi-custom design is an exploitation of basic cell libraries that may not be changed by an user. A more flexible approach is achievable by means of 'silicon compilation' actually being introduced into the ASIC marketplace.

Analog integrated circuit design unveils different requirements compared to its digital counterpart. The most important requirement is flexibility even in an array based design environment. Consequently, a design should exploit as many library cells as possible to cope with tight design time limits, but it needs to exploit custom designed cells in order to handle properly system specific requirements. The ability of the design system to cover the electrical design of high-performance cells is crucial to the performance of an entire circuit. An important condition for a successful design of high-performance cells is an accurate and reliable simulation. This, in turn, requires powerful tools for device and cell characterization. Another different requirement compared to digital circuit design is the versatility of the tool aimed to performance assessment. Finally, the layout generation tool of an analog circuit design system requires specific features to achieve first time functional designs in presence of a still limited applicability of design automation methods.

The tool set presented in the following is subjected to on-going development and improvement according to the fast changing requirements present in the semiconductor business. Its present state may be viewed as a 'design system for (human) experts'. However, some fundamental ideas for a knowledge based tool are outlined in the last section of this chapter. This concept is a first step towards an analog design 'expert system' in its conventional meaning.

# 2 OVERVIEW OF A HIERARCHICAL ENGINEERING SYSTEM

The structure of the engineering system presented in the following sections is outlined in Fig. 1. It consists of a central data base and of tools dedicated to the tasks of electrical and physical design. The data base incorporates the schematic, electrical and physical views of each cell as well as technology related data. A cell may contain a single transistor, a subcircuit or an entire circuit. Views of an entity correspond at each

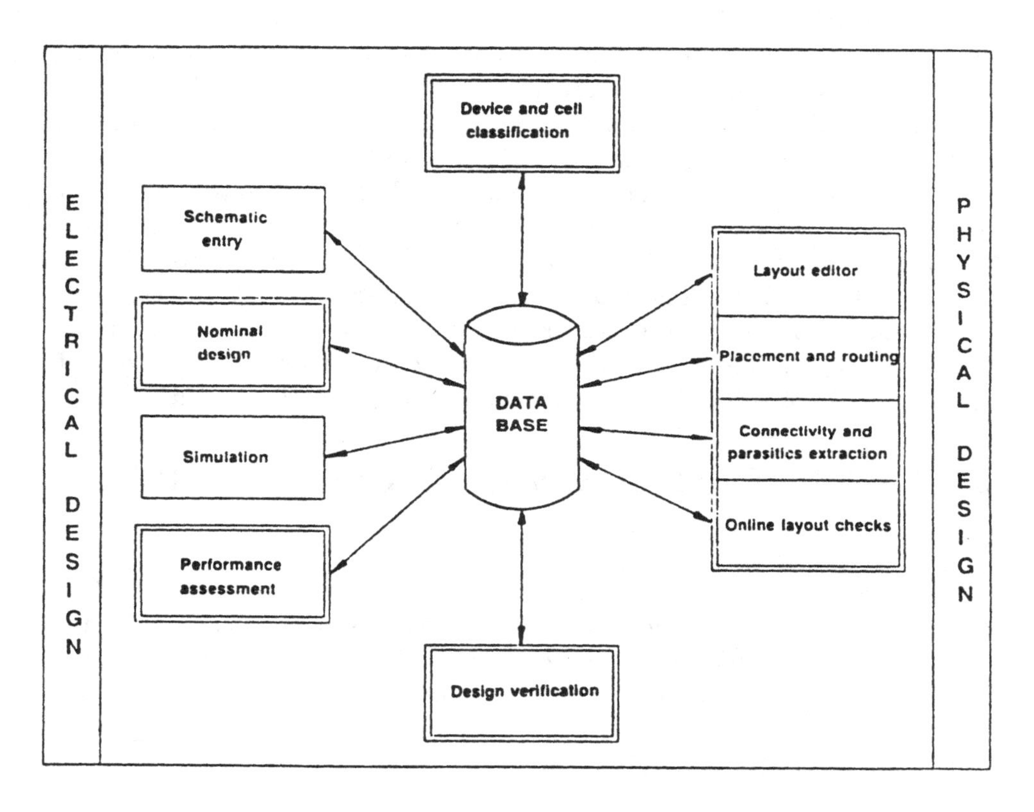


Figure 1: Structure of the engineering system.

level in hierarchy.

The implementation of the data base exploits the file structure of the Unix based hardware platform of the engineering system. An application of a commercial relational DBMS has been considered, but experimental results unveiled an insufficient performance of the entire system in terms of data access speed. However, the outcome of the data modeling activity performed with respect to an application of a relational DBMS was quite valuable for an efficient implementation of the data base on top of the file management system.

The electrical design part consists of two characterization packages, of a schematic entry subsystem, of simulation and performance assessment tools as well as of an optimization based nominal design subsystem aimed to high-performance cell design.

The physical design part incorporates an interactive layout generation package with embedded connectivity and electrical rule checks as well as an integrated extraction facility dedicated to physical layout parasitics. A design verification subsystem is present, which consists of two packages. The first one is aimed to netlist and component value comparison between the validated schematic and the associated layout of a circuit. The second package is a sophisticated design rule checker, which may be invoked either for an off-line check after the completion of a layout or in an on-line mode during an interactive layout session.

These tools are proprietary except the schematic entry and simulation packages. Therefore, they are streamlined to cover a variety of special requirements of high-frequency, non-linear analog circuit design.

# 3 DEVICE AND CELL CHARACTERIZATION

The characterization part of the system consists of two packages. The first one is aimed to device characterization based on measured data. The second package is dedicated to an exploration of the function space of analog cells by means of circuit level simulation.

## 3.1 Device characterization

This tool covers the area of parameter extraction for simulation models by fitting parameters of model equations to measured data. In addition, it is used for process monitoring tasks. Fig. 2 shows the simplified data flow within the tool.

Measurements are performed in both AC and DC domain yielding the following objectives.

- AC :

    1. C-V data base-emitter, base-collector and collector-substrate capacitance values
    2. S-parameters
    3. base resistor values by Time Domain Reflectometry

- DC :

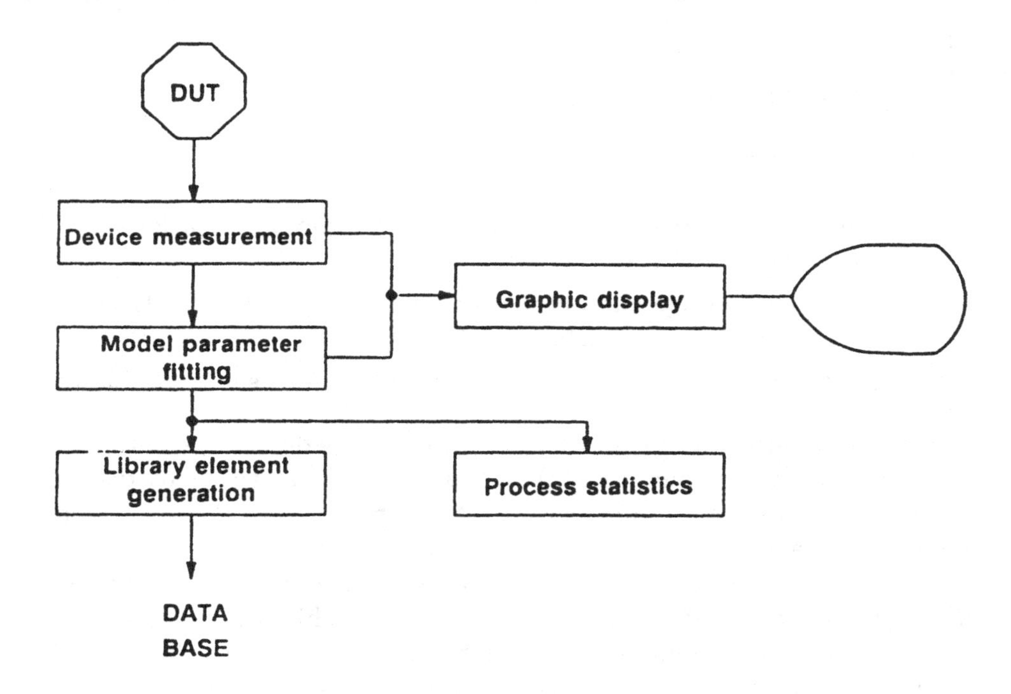


Figure 2: Data flow for device characterization.

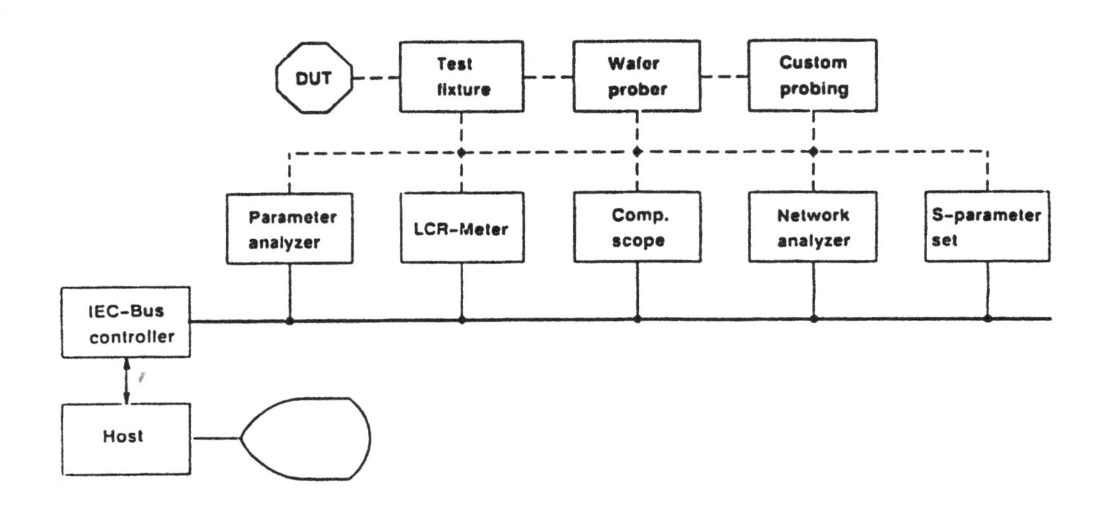


Figure 3: Hardware configuration for device characterization.

1. current-voltage characteristics for bipolar and MOS transistors, diodes and resistors
2. process data such as breakdown voltage and sheet resistance

Fig. 3 outlines the hardware configuration of the device characterization tool. The device under test (DUT) can be accessed either by means of a test fixture, a wafer prober or a custom probing unit depending on its physical shape.

The measurement units are controlled via an IEC bus by the characterization software running on a host computer. This software package consists of three main modules. The first one is dedicated to the control of the measurement equipment. Its output consists of the measured data for the DUT. The second module performs the fitting of the simulation model parameters to the measured data. This task is subdivided into two phases. In the first phase a statistical optimization method is applied for a first estimation of the model parameter values. The second phase exploits a deterministic optimization method for the refinement of the estimated parameter values. The output of this module is a parameter set for the selected simulation model. The parameter sets are then inserted into the central data base of the engineering system. The purpose of the third module is statistics generation for process monitoring.

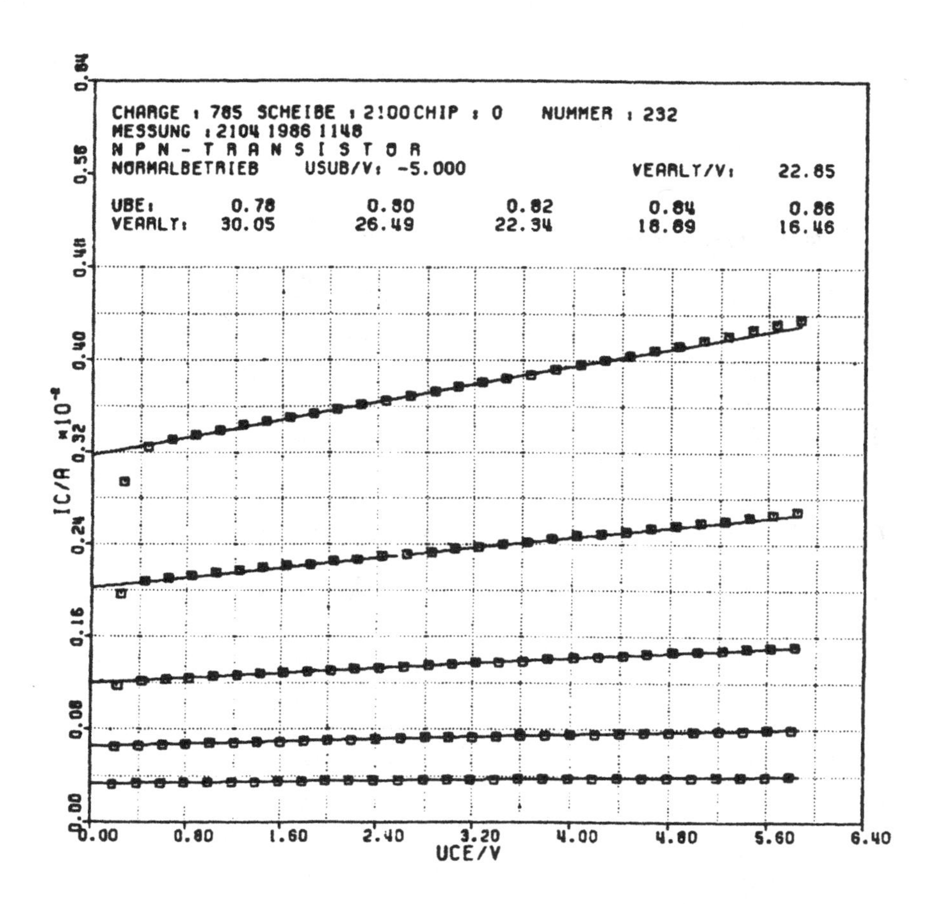


Figure 4: DC output characteristics of a UHF transistor.

A more detailed description of this characterization tool may be found in [1]. Some results gained from npn-type transistors are presented in the following.

Fig. 4 depicts the plot of the DC output characteristics of a UHF npn-type transistor, whereas the symbols visualize measured values and the curves depict the results of the simulator model for the calculated parameter set. The estimation of transit frequency values is outlined in Fig. 5 yielding a value of about 5 GHz for the DUT.

The accuracy of model parameters is crucial to a reliable simulation and hence for a successful design of high-frequency circuits. The

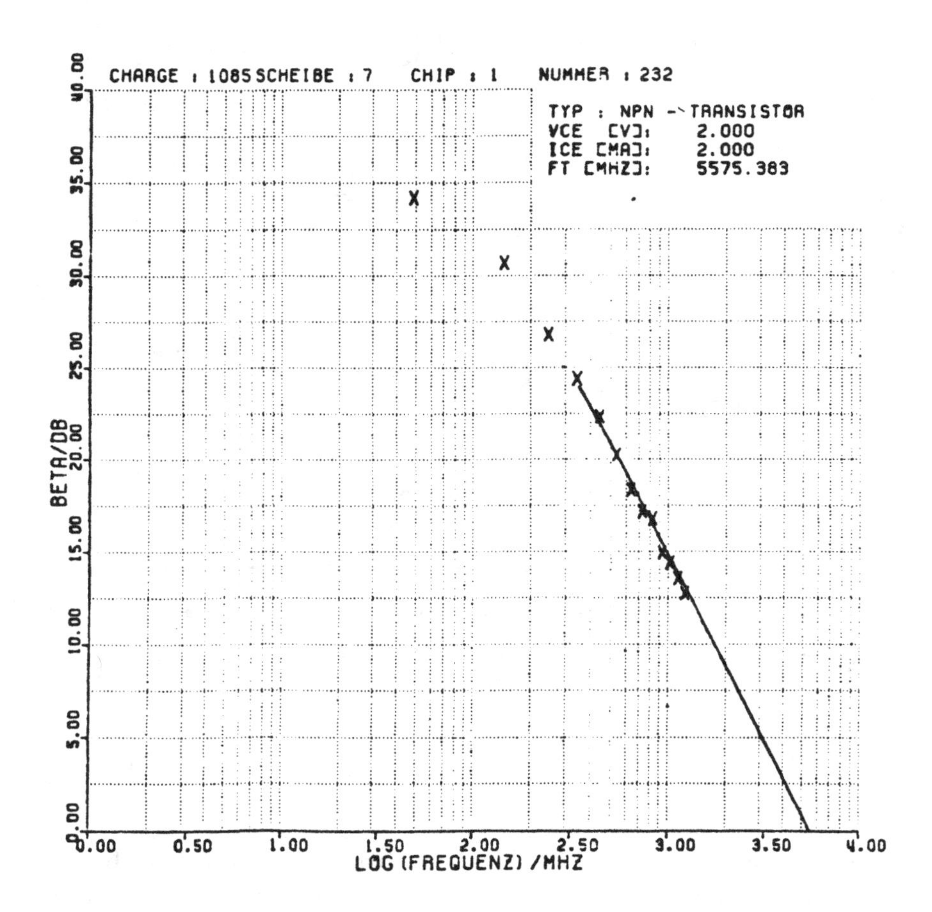


Figure 5: Estimation of transit frequency value.

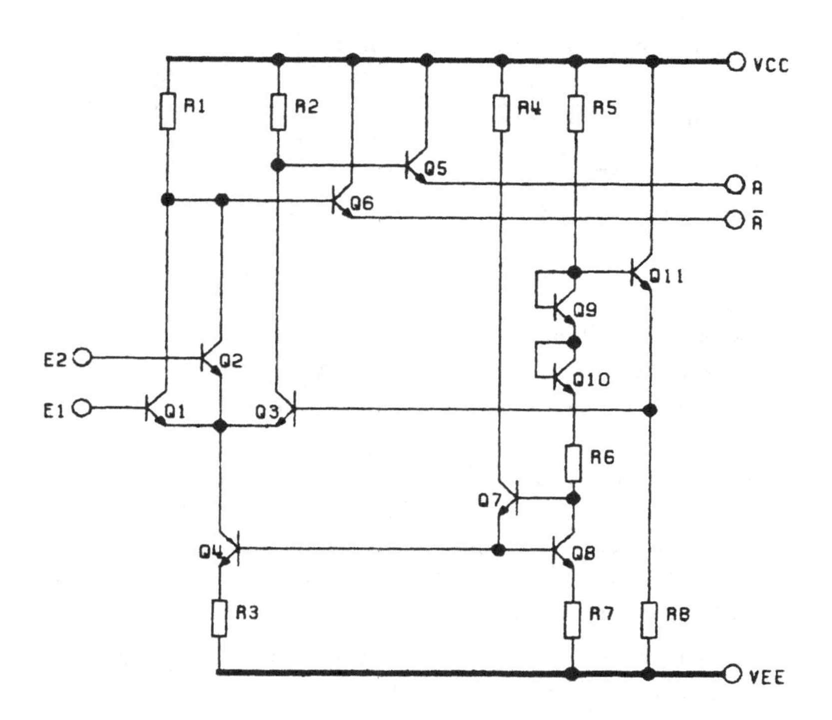


Figure 6: Schematic of a two input NOR gate.

achievable accuracy of performance prediction is demonstrated for the time domain analysis of an ECL circuit. Fig. 6 shows the schematic of a NOR gate. Table 1 summaries delay and waveform characteristics of this circuit gained from both simulation and measurements. Deviations are within a 10accuracy of this device characterization tool.

### 3.2 Cell characterization

The second characterization package is aimed to an exploration of the functional space of analog cells such as amplifiers, mixers or voltage reference circuits. Its output is a set of data sheets, which are intended to support a fast and appropriate selection of cells for a specific design application.

| Objective | Simulation | Measurement |
|---|---|---|
| Propagation delay | 410 ps | 400 ps |
| Rise time 10%-90% | 630 ps | 610 ps |
| 20%-80% | 390 ps | 380 ps |
| Fall time 10%-90% | 540 ps | 510 ps |
| 20%-80% | 280 ps | 300 ps |

Table 1: Comparison of simulated and measured results

This package is driven by a circuit level description of the cell to be characterized and a command file. The command file consists of three main sections, which contain property, parameter and condition definitions, respectively.

The first section contains analysis domain definitions and the objectives of each domain such as phase margin in the frequency domain. In the second section design parameters and their feasible ranges are given. Finally, the third section is dedicated to condition specifications, which are to be met for a functional circuit.

The cell characterization package produces automatically for each feasible parameter combination out of the hypercube associated to the given parameter ranges an input deck for the simulation package, invokes the simulator, extracts the values of the specified properties, checks the given conditions and produces after completion of these tasks a set of data sheets. In addition, sensitivity calculation can be performed for a given nominal design of the circuit.

The mixer circuit shown in Fig. 7 was characterized for a total of ten properties (e.g. output resistance, noise figure at a given frequency point and multiplication value) depending on three design parameters (reference current and two resistor values), whereas no conditions were specified.

Fig. 8 shows one of the data sheets produced by the characterization tool. The range of each property depending on the parameter range definitions are summarized in Fig. 8. For example, the values of the bandwidth denoted as FG range from 4.59 MHz up to 289 MHz. Functional space exploration as implemented in this characterization package allows for an easy selection of suited cells for a particular design application.

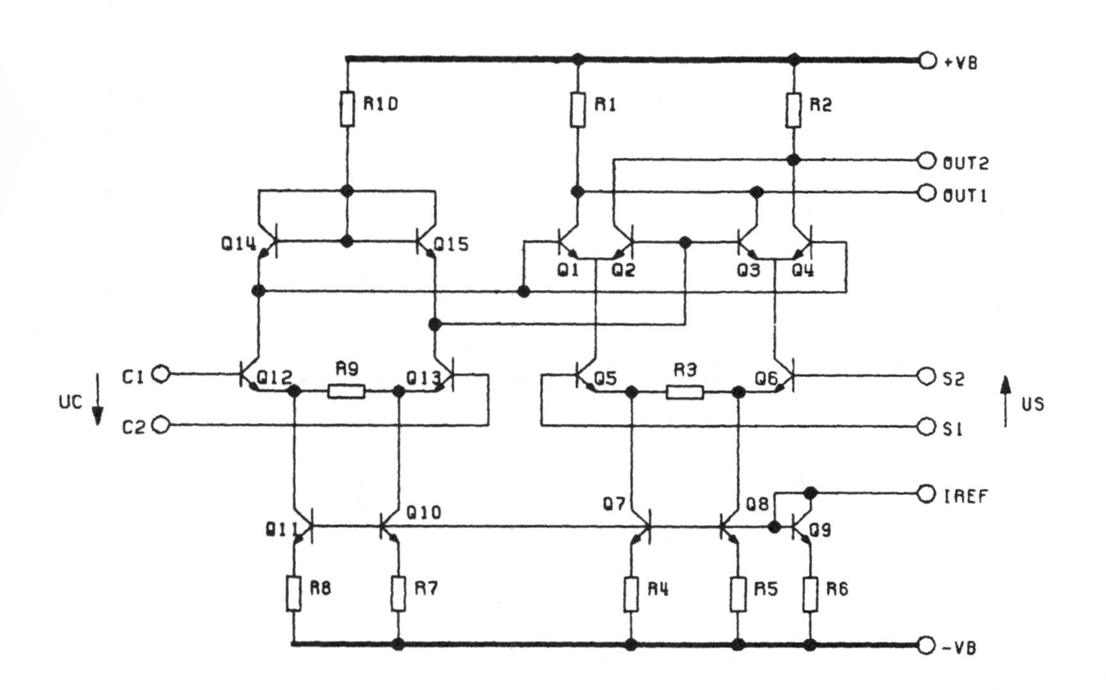


Figure 7: Schematic of a mixer circuit.

AEG AKTIENGESELLSCHAFT
DESIGN CENTER FOR INTEGRATED CIRCUITS

| PARAMETERS | | | | | |
|---|---|---|---|---|---|
| NAME | MIN | MAX | NAME | MIN | MAX |
| IREF | 1.00E-04 | 1.00E-03 A | R1 (=R2) | 2.00E+02 | 2.00E+04 OHM |
| R3 | 2.00E+02 | 2.00E+04 OHM | | | |

| PROPERTIES | | | | | |
|---|---|---|---|---|---|
| NAME | MIN | MAX | NAME | MIN | MAX |
| IES | -7.13E-06 | -7.58E-07 A | ICC | -4.83E-03 | -4.82E-04 A |
| M | 9.43E-02 | 9.08E+01 1/V | RS | 3.50E+04 | 2.72E+06 OHM |
| RAS | 3.99E+02 | 3.94E+04 OHM | RAU | 2.00E+02 | 1.99E+04 OHM |
| ENOS | 3.38E-09 | 3.70E-07 V/SQR(HZ) | ENOU | 2.32E-09 | 2.34E-07 V/SQR(HZ) |
| FG | 4.59E+06 | 2.89E+08 HZ | PHFG | -2.06E+02 | -3.14E+01 GRAD |

CELL : SKCMU1

DATE :25-JAN-90 PROCESS :BIP2
USER :RUFF440 SIMULATOR :SPICE2G6
AEG FI4-UL PAGE 1 OF 7 CELLPLOT : APL REV. 1.2

Figure 8: Characterization results for the mixer circuit.

# 4 ELECTRICAL CELL DESIGN

Simulation has meanwhile become a standard method in the area of electrical cell design at transistor level, which allows for a fast and reliable assessment of the effects on cell performance resulting either from different circuit topologies or from nominal design variants of a fixed topology. The selection of possible nominal designs is usually based upon approximations and the designer's experience. Beyond this state of the art, a need is evident for systematical and efficient methods for nominal design generation in order to meet performance specifications [2,3]. The nominal design problem is defined as follows: Determine feasible values of designable parameters of a circuit given by its topology such that the requirements of circuit properties are fulfilled. This task is usually transformed into a multidimensional, non-linear optimization problem, which involves multiple and often conflicting criteria.

The nominal design tool of this engineering system is based on the least squares optimization criterion and requires first-order sensitivity (gradient) and circuit performance data. It exploits the cell characterization package outlined in the previous section to produce these data. A short description of the approach implemented for nominal design is given in the following.

Denoting f the vector of circuit properties depending on the vector p of designable parameters, an error vector $\varepsilon$ after any correction x of the actual values of p is obtained from

$$\varepsilon(x) = f(p + x) - f_s \tag{1}$$

whereas $f_s$ denotes the vector of circuit specification values and x denotes the vector of parameter changes. By linearizing the normally non-linear error vector, the least squares objective function is given by

$$\phi(x) = ||\varepsilon(x)||^2 = ||\varepsilon_f + Sx||^2 \tag{2}$$

and solved by an iterative process, whereas S denotes the Jacobian matrix. The stationary point $x_0$ of $\phi(x)$ results from the Newton step. The error vector after the correction of p by $x_0$ is denoted as $\varepsilon_0$. The modification of the objective function by a Lagrange multiplier results in the Levenberg-Marquart formulation of minimizing $\phi(x)$ within a hypersphere.

$$\phi(x, \lambda) = ||\varepsilon(x)||^2 + \lambda||x||^2 \tag{3}$$

Denoting the normalized $l_2$ norm of $x_0(\lambda)$ as correction effort $a_0(\lambda)$ and the transformed $l_2$ norm of $\varepsilon(\lambda)$ as a measure for the associated correction effect $r_0(\lambda)$, we obtain by varying $\lambda$ a boundary curve between effect and effort in a parametric representation

$$r_0(\lambda) = g[a_0(\lambda)] \tag{4}$$

which is characteristic to the optimization problem at hand. This curve is best suited as an assessment criterion for the interactive control of the entire nominal design process and forms therefore the basis of this electrical design tool.

The interactive optimization strategy considered here consists of two stages and features a novel decision making method [4]. In the first stage, the dominant subset of parameters and/or circuit properties is established resulting in a substantial saving of computing effort during the next iteration cycles and in an improvement of the numerical condition of the optimization task. The final correction of the parameters is then determined in the second stage.

This electrical design tool provides the following main features [5]:

- automatic generation of performance and sensitivity data using circuit level simulation
- generation of suited nominal design proposals under user control
- on-line visualisation of the predicted performance improvements.

The first and third item are covered by the cell characterization tool described previously. The second item is handled by the interactive optimization package outlined above. Its application is demonstrated for the electrical design of an integrated broadband amplifier [6] aimed to radar signal preprocessing. The schematic of the amplifier is shown in Fig. 9 and the design specifications are given in Table 2.

The first optimization step is outlined in Fig. 10 to 12, which are photographs of the workstation screen. The background window of Fig. 10 contains a printout of the command file for performance and sensitivity data generation.

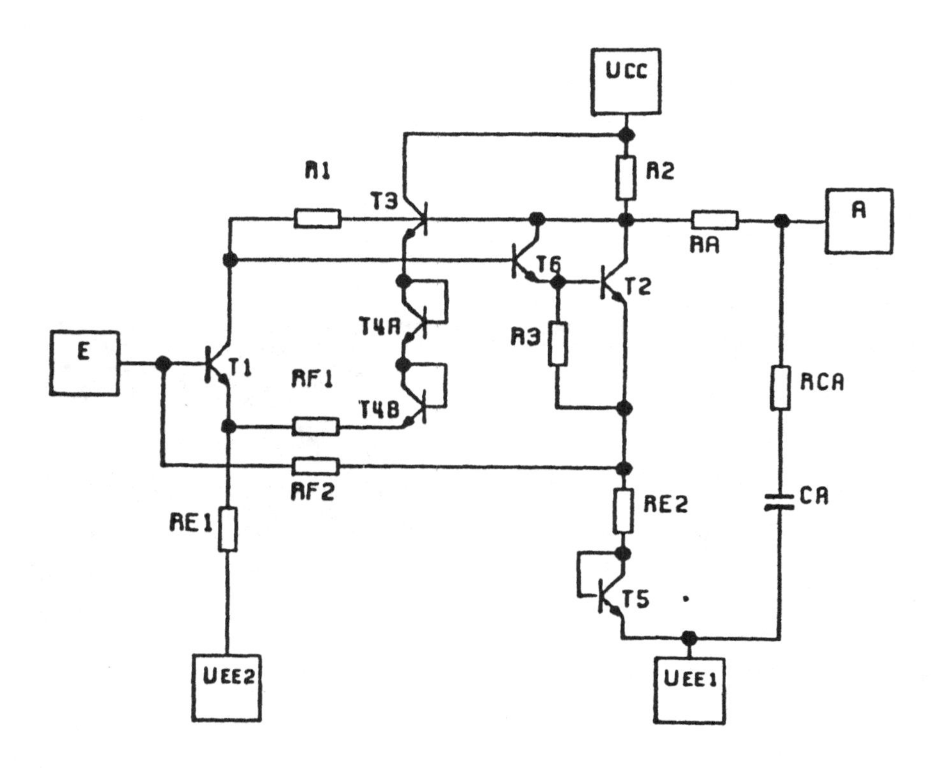


Figure 9: Schematic of a broadband amplifier.

| Objective | Value |
|---|---|
| Gain | > 10 dB |
| Bandwidth | > 2.5 GHz |
| S-parameter s11 | < 0.2 |
| S-parameter s22 | < 0.2 |
| Input noise | $< 2\ \mathrm{nV}/\sqrt{Hz}$ |

Table 2: Design specifications of the amplifier.

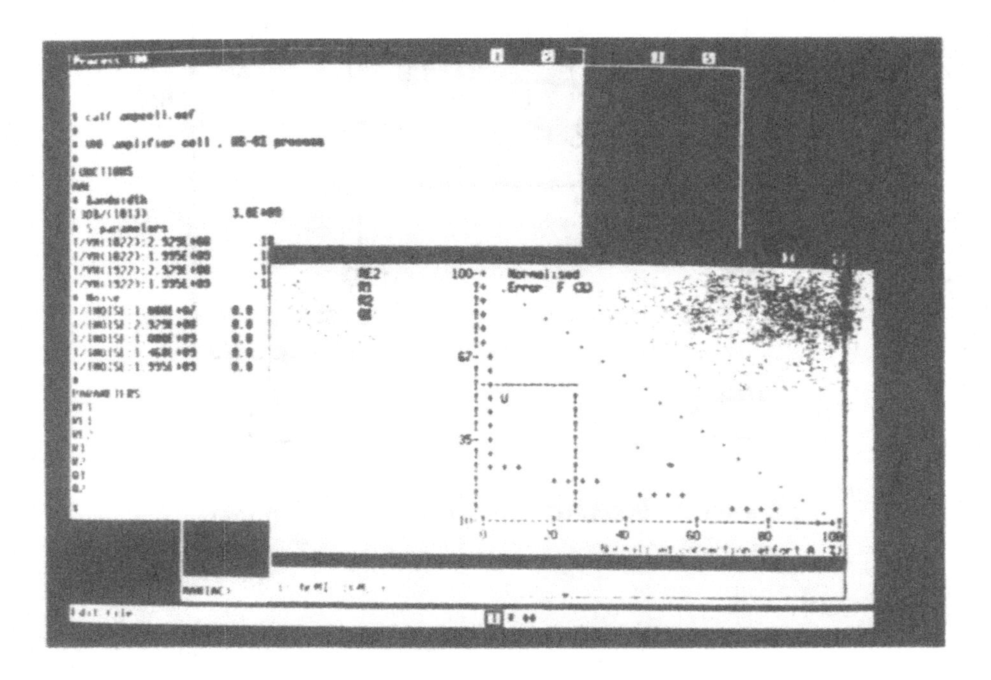


Figure 10: Selection of the relevant parameter subset.

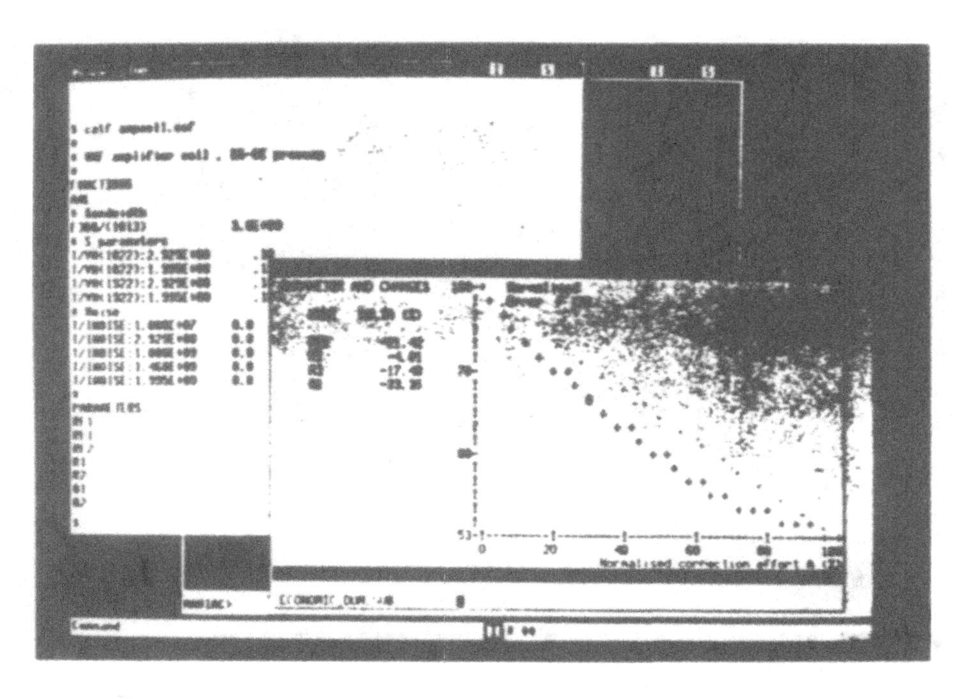

Figure 11: Calculation of the parameter values.

In the first section the circuit functions or properties are defined, whereas the second section contains the names of seven designable parameters. The foreground window depicts the boundary curve. By means of this assessment criterion a relevant parameter subset was established automatically according to the user defined target area near the sharp bend of the curve, which is characteristic for parameter redundant (i. e. ill-conditioned) problems [4]. The parameter subset found contains four parameters. The position of the subset is marked within the target area.

Now this subsystem is used for the calculation of the assessment criterion as shown in Fig. 11. Obviously, there is no more sharp bend as for the entire parameter set. This visualizes the considerable improvement of the problem condition. The solution marked on the curve and denoted by the parameter changes visualized in the left upper part of the window is now verified by performing a simulation run for the changed

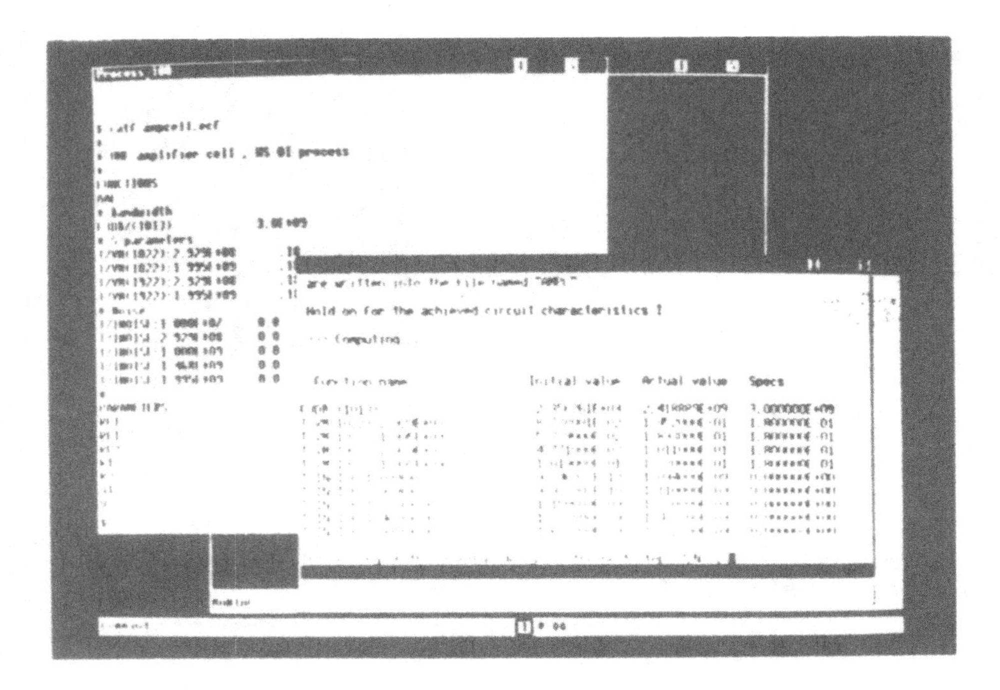


Figure 12: Verification of the new nominal design.

parameter values without leaving the optimizer. Its results are given in Fig. 12.

The initial, actual and specified values are summarized for each property given in the command file. For instance, the effect of this first optimization step increased the bandwidth from 2.36 GHz to 2.42 GHZ as visible from Fig. 12. After two more iterations the circuit met the specifications. The bandwidth has been improved by 15features 2.63 GHz for the final version of the circuit.

# 5 PERFORMANCE ASSESSMENT

The tool aimed to performance assessment was designed for versatility and flexibility demands present in analog circuit design. First, result representation familiar to design engineers and a wealth of data manipulation facilities are mandatory for a fast and error-free assessment of circuit properties. This requirement goes far beyond of just generating some x-y plots from simulated waveforms. Second, comparison operations between results gained from different simulation runs or from runs of different simulators are needed as well as comparisons between simulated and measured data. The support of a variety of data sources is aimed to an exploitation of different simulation packages because of specific strengths and weaknesses of each simulator for a particular application.

The first requirement was met by providing following data representation and manipulation facilities:

- multiple variable plots
- Bode, eye and locus diagrams
- Smith, polar and spectrum charts
- mathematical operations including complex data calculation, FFT and integration
- automatic extraction of circuit properties.

The second requirement was considered by implementing simulator specific data loading modules for at this time three general purpose circuit analysis packages, for two switched-capacitor analysis programs, for the device characterization tool outlined above as well as for a general data format aimed to measured circuit property values.

Some examples of typical data representations are shown in the following figures, whereas the simulation results were produced using Spice. Fig. 13 shows the Bode diagram of a preamplifier, whereas the extracted bandwidth is marked by a symbol. Fig. 14 outlines the eye diagram for a bipolar crosspoint module [7] at a data rate of 400 Mb/s.

Fig. 15 visualizes time domain results for a frequency doubling circuit, which in turn is part of an IQ-demodulator. The input signal is the pulse at 500 MHz and the output signal is the 1 GHz sine-like curve of Fig. 15. Fig. 16 depicts the calculated spectrum of the output signal.

The data representation examples are concluded by the Smith chart of Fig. 17, which depicts simulated (solid line) and measured (symbols) values of scattering parameters in a frequency range from 50 MHz up to 1.3 GHz. A more detailed description of this interactive analog waveform analyses is given elsewhere [8].

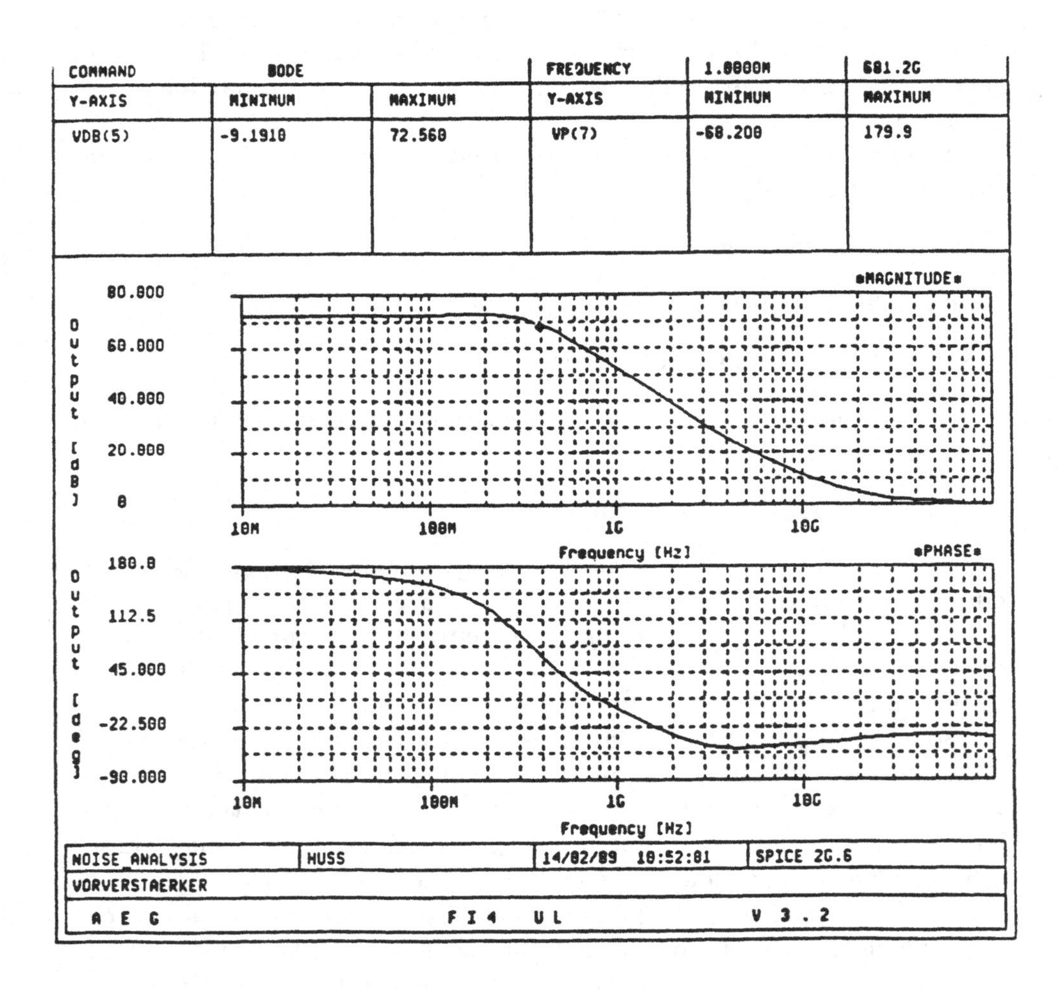


Figure 13: Frequency domain response of a preamplifier.

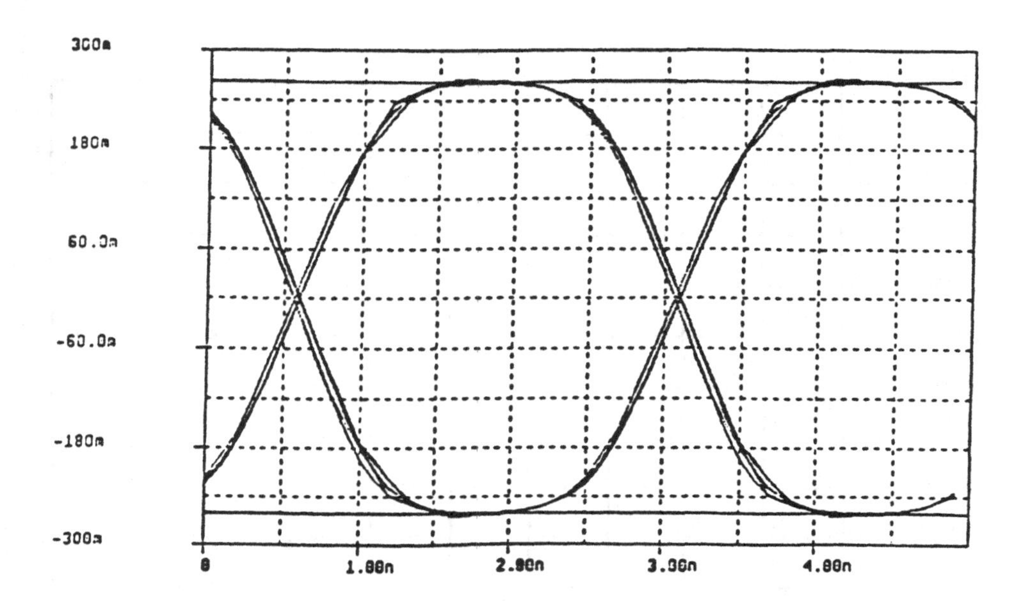


Figure 14: Eye diagram from output signals of a crosspoint module.

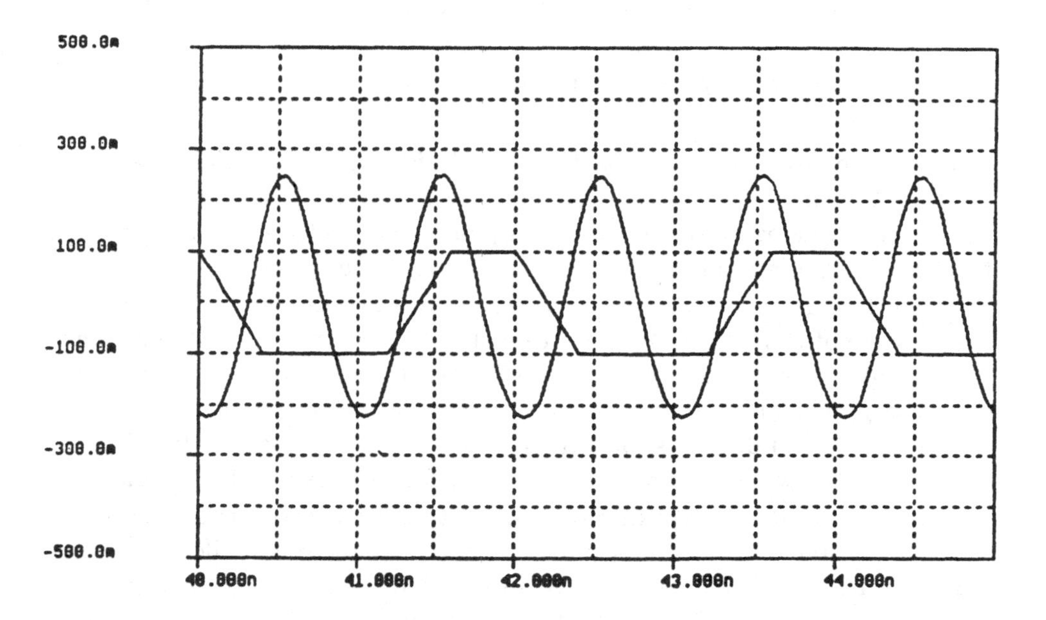


Figure 15: Time domain results for a frequency doubling circuit.

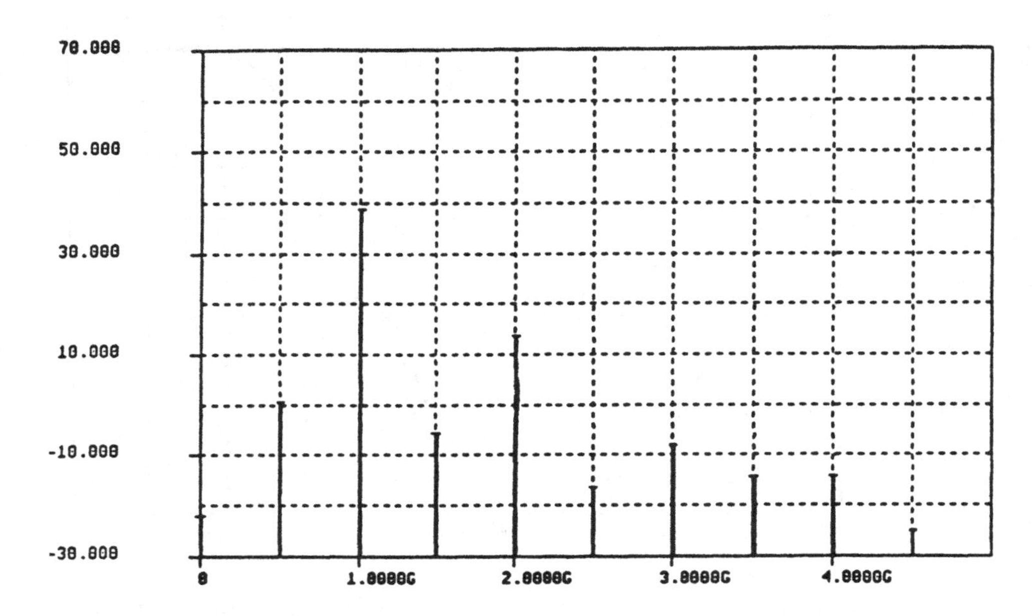


Figure 16: Associated spectrum of the output signal.

# 6 PHYSICAL DESIGN

A layout generation tool is part of the design system, which has been developed with respect to requirements present in both semi-custom and full-custom analog circuit design. Emphasis was put on layout verification in an on-line mode and on layout parasitics extraction.

This tool is based on a sophisticated data structure that allows for the storage of not only geometrical mask data but also connectivity and functional attributes of layout components. This program internal data base is organized hierarchically according to the circuit structure. In addition, the layout data is kept during an interactive design station in the memory of the workstation yielding a very fast access to all components.

The layout editor working on top of this data structure allows for real time connectivity checks and extraction operations while executing edit commands. Their results are inserted immediately into the data structure yielding actual connectivity information at any design step. The key features of this layout generation tool are summarized in the following:

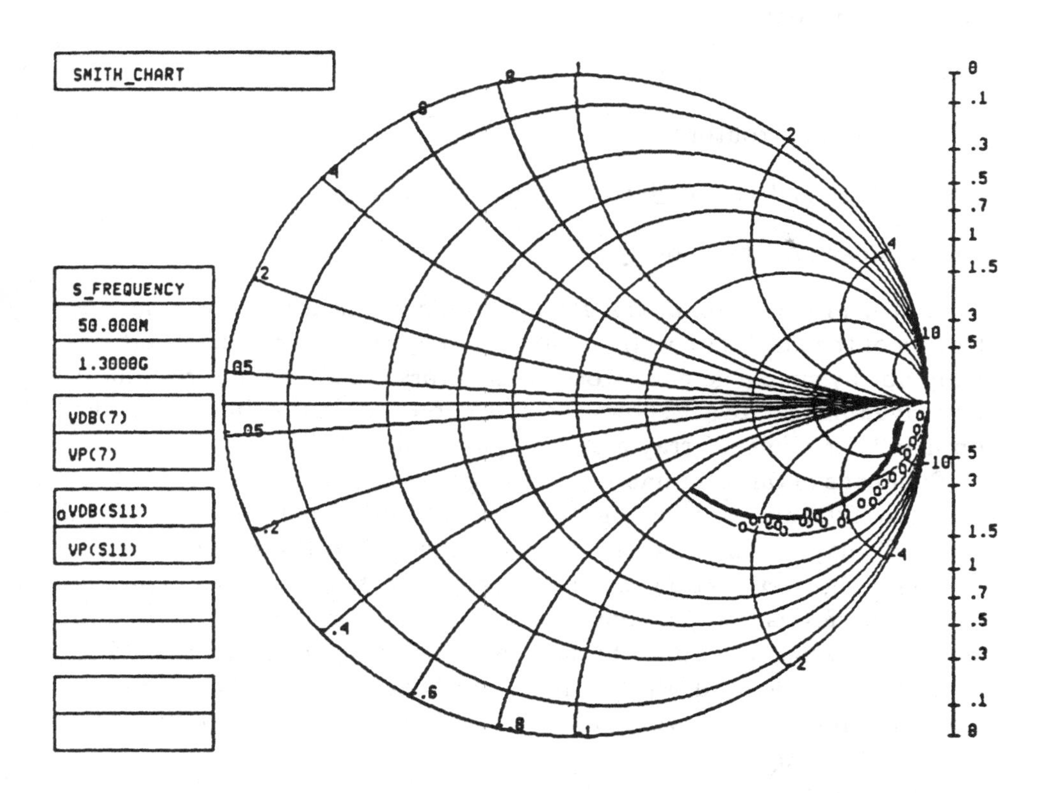


Figure 17: Comparison of measured and simulated data.

- hierarchical data structure containing geometrical data and functional attributes of components
- extraction of circuit connectivity and layout parasitics including resistors, two- and three-dimensional capacitances from the routing
- basic cell layout generators
- automatic cell placement
- single net auto-routing
- Manhattan and user defined component shapes
- independence of technology.

A detailed presentation of this layout tool is given in [9]. For a demonstration of the kind of operations supported by this layout editor but certainly not by simple geometry oriented editors consider the screen hardcopy given in Fig. 18. It shows the screen output of the "inspect_node" command for a linear array design application.

By means of this command all routing elements of the selected net are immediately found and highlighted. In addition, a list of all contacted cells is displayed including their ports connected to the selected net.

The design verification task is addressed by two packages. The first package performs comparisons of netlists and component values produced from schematic and physical representations of a circuit. These checks may be invoked directly from the layout editor. They are executed hierarchically according to the corresponding cell hierarchy within electrical and physical design.

The second package is aimed to layout auditing for full-custom design applications, since array and standard cell based designs are 'correct by construction' with respect to layout design rules. Technology rule definitions are loaded directly into the layout generation tool and are then considered accordingly in order to ensure a correct design.

The scope of this design verification tool goes far beyond of merely checking distances or widths of geometrical components. It can be viewed as a user programmable mask data manipulation tool. It operates on non-Manhattan entities featuring any angle value and allows for a variety of manipulations of entities depending on their local environment by means of user definable operations (e.g. Boolean operations)

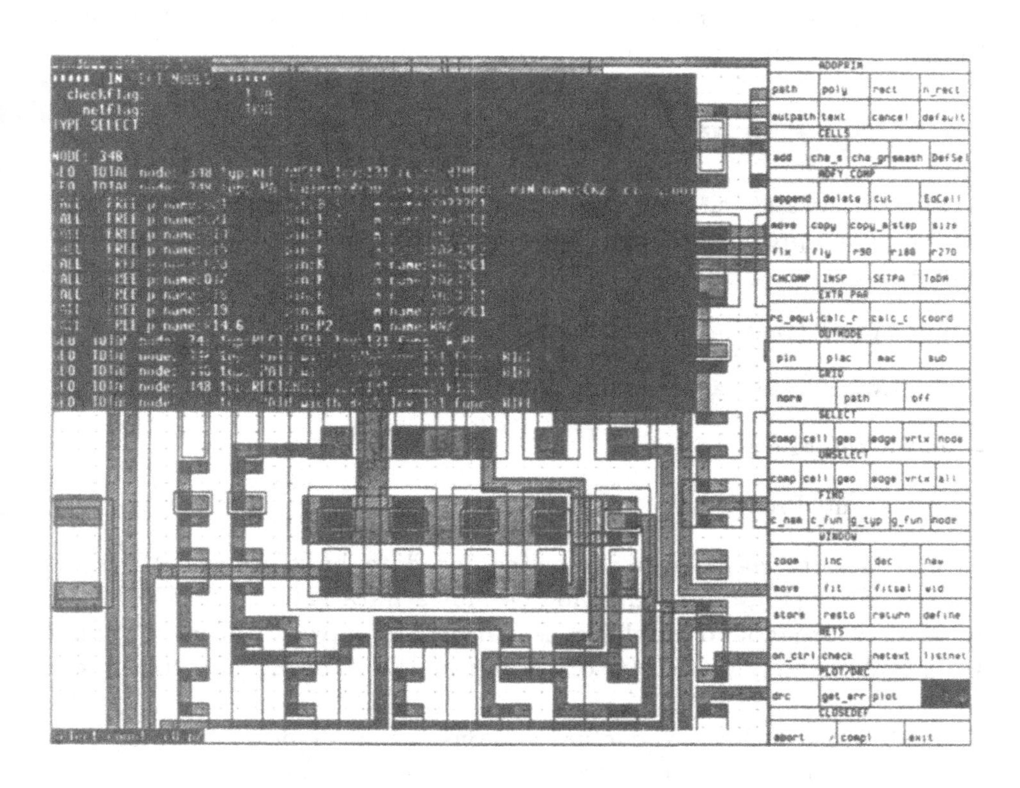


Figure 18: Screen representation of the layout editor.

on layers and by individual sizing of components or layers. Abutment and enclosure checks are therefore very easy to perform.

The verification package can be invoked during an interactive layout design session. It operates then in an user defined layout window and allows for an immediate backannotion of errors into the graphic editor. The powerful checking facilities of this package and its easy handling are of special interest in presence of somewhat odd structures not uncommon to high-frequency bipolar circuit layouts.

# 7 DESIGN EXAMPLES

Analog integrated circuits supported by the engineering system presented in the previous sections are aimed to the following main application areas:

- high-speed transmission systems
- preprocessing of radar signals
- sensor signal conditioning
- broadband ISDN.

Circuit designs have been completed for data links from 2 Mb/s up to 565 Mb/s [10], crosspoint modules for a data rate up to 565 Mb/s, integrated broadband amplifiers featuring a bandwidth of more than 2 GHz and homodyn receivers operating in the X-band, just to name a few.

The tools outlined above are demonstrated for designs aimed to a standard bipolar technology for 'low frequency' applications and in an UHF technology for applications up to 3 GHz.

This technology features fully implanted, oxide isolated devices. The achievable transit frequency figure is about 7 GHz. Semi-custom as well as full-custom designs are supported for both technologies. A family of linear arrays containing 250, 500 and 1000 components, respectively, and an analog standard cell library are available for semi-custom designs using the standard technology. A linear array featuring 4000 components is available for the UHF technology. The components of this UHF linear array are arranged into four blocks, which may be used for flexibility purposes either on a per block or per chip basis.

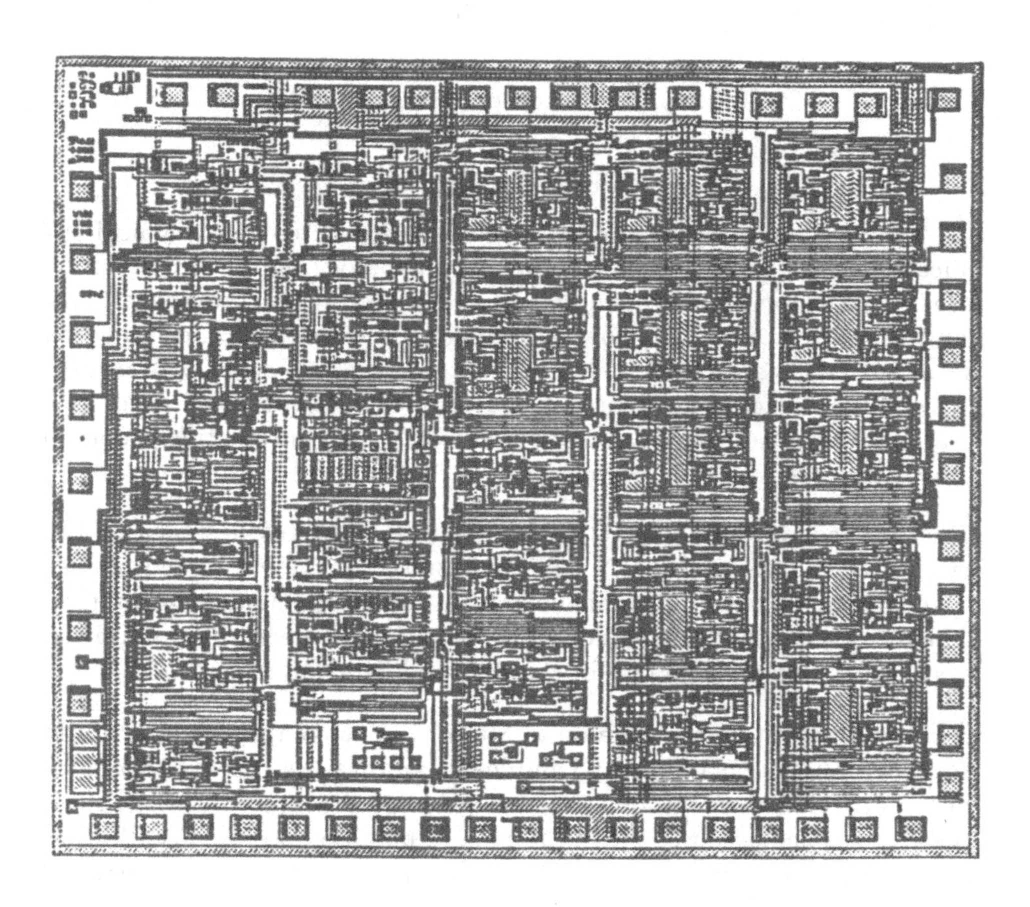

Figure 19: Layout of a standard cell analog/digital circuit.

| Objective | Transmit | Receive |
|---|---|---|
| Bit rate | 280 Mb/s | 280 Mb/s |
| Input signal | 20 mVpp | |
| Modulation current | 84 mA | |
| Bandwidth | | 150 MHz |
| Sensitivity (at a binary error rate of 1e-09) | | -33 dBm |

Table 3: Measured performance data.

Two design examples are presented in order to highlight typical applications. The first example is a mixed analog/digital circuit to be used in an automation system and has been produced in standard bipolar technology. Fig. 19 shows its layout. The cell placement of this standard cell based circuit was done automatically by means of a force directed cell placement algorithm [11] implemented in the layout generation tool, whereas the routing was performed interactively due to crosstalk sensitive signal paths.

The second example is a 280 Mb/s optical data link designed on top of the UHF linear array [12]. The transmission system is digital, but its components need to be engineered by means of analog techniques. This situation is not uncommon in designing high-speed communication systems. The block diagram of the data link is given in Fig. 20.

The transmit and receive units are of similar complexity. Each unit consists of nine cells of opamp complexity and was designed into a block of the array with a minimum amount of external components. All these cells except the preamplifier of the receive unit are part of a cell library associated to this linear array. The performance of the preamplifier cell regarding bandwidth and noise is significant to the overall performance of the data link. Its electrical design was completed by means of the optimization based nominal design tool resulting in a customized cell specific to this application.

The cell based structure of the circuit is well visible from the layout of the receive unit as depicted in Fig. 21. Some measured performance data of this optical data link are summarized in Table 3.

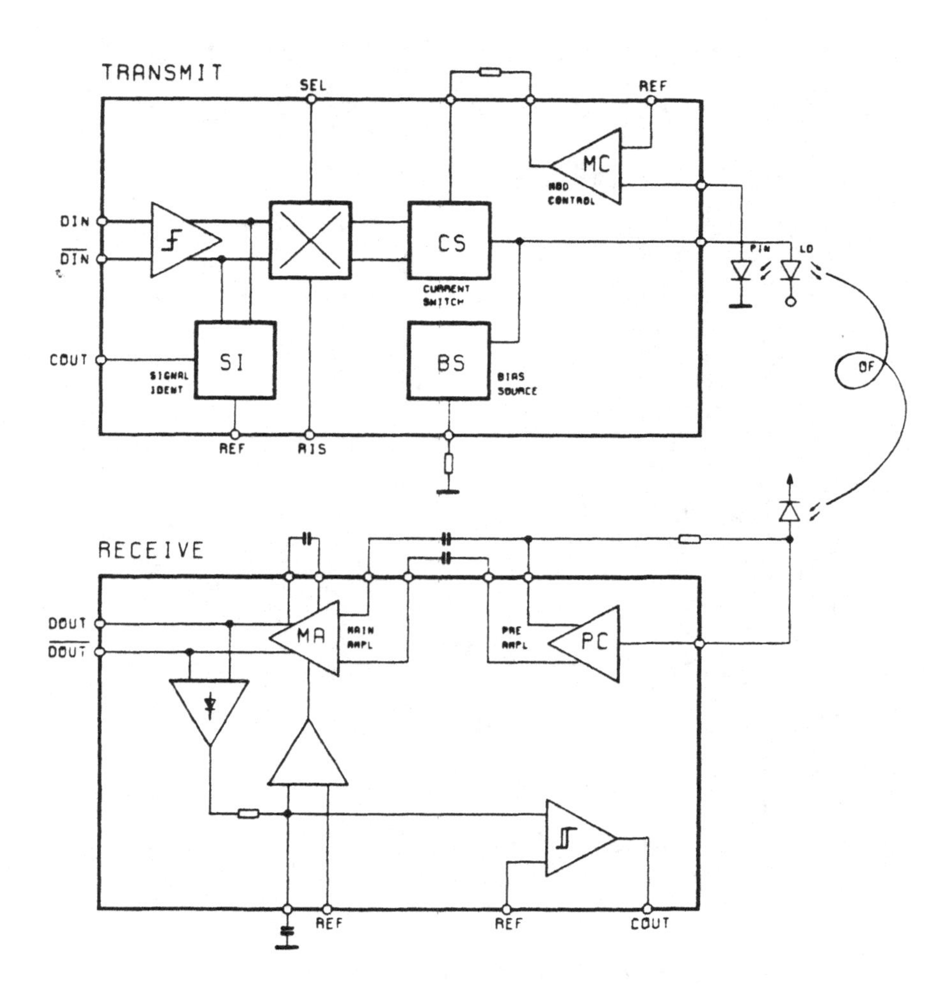


Figure 20: Block diagram of a 280 Mb/s optical data link.

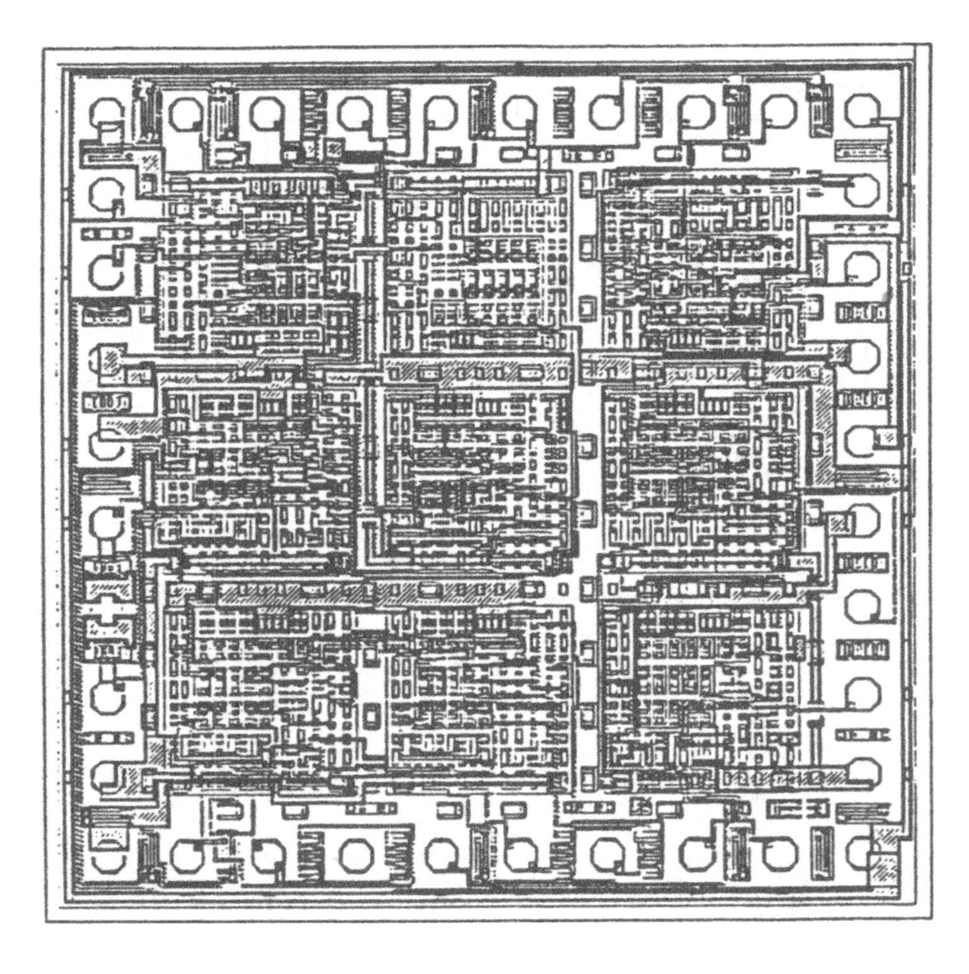

Figure 21: Layout of the receive unit.

# 8 RULE BASED DESIGN ASSISTANT

Analog circuit design is still a task for experienced designers. On the other hand, an increasing demand for application specific analog circuitry is present in the marketplace. As an attempt to improve this situation, a workpackage was initialized by four companies and institutions as part of the ESPRIT project entitled "CAD - VLSI for Systems". The main goal of this workpackage is to enable novice and little experienced circuit designers to cope with analog design tasks by means of an engineering environment acting as a design assistant.

Knowledge about how to construct analog circuits is an essential part of this engineering environment. The approach is based on hierarchical 'skeleton cells' and on rules associated to them. A set of rules is available for each skeleton, which may be a transistor, a subcircuit consisting of transistor skeletons, or a cell assembled from subcircuit skeletons. A skeleton at a higher level in hierarchy is thus constructed from lower level skeletons by means of rules according to given specifications. A first introduction to this concept is stated in [13]. The work towards a knowledge based engineering environment is actually well in progress, but not yet completed.

This rule based approach to cell construction by selecting appropriate skeleton implementations is demonstrated for the assembly of an operational amplifier. The skeleton representation of an opamp is depicted in Fig. 22.

There are a set of implementations and a set of rules stored in the library for each skeleton labeled A to G in Fig. 22. The rules are considered during the selection of an implementation of the skeleton to meet given specifications. Subsets of rules associated to input stage and load stage skeletons of an opamp are given in Tables 4 and 5, respectively. Rules are structured into three parts: objective, condition and action.

The available set of skeleton implementations for an opamp input stage is outlined in Fig. 23.

A total of nine implementations are selectable, which may differ just in the type of the input transistors as for DS1 and DS2, or unveil a completely different connectivity as for DS4 and DS8 of Fig. 23. How the selection process works is shown by means of the second rule of Table 4. In case that the specified input noise current (objective) is less than a technology dependent threshold value (condition) then the npn-type Darlington stage denoted as DS3 in Fig. 23 is selected (action) instead

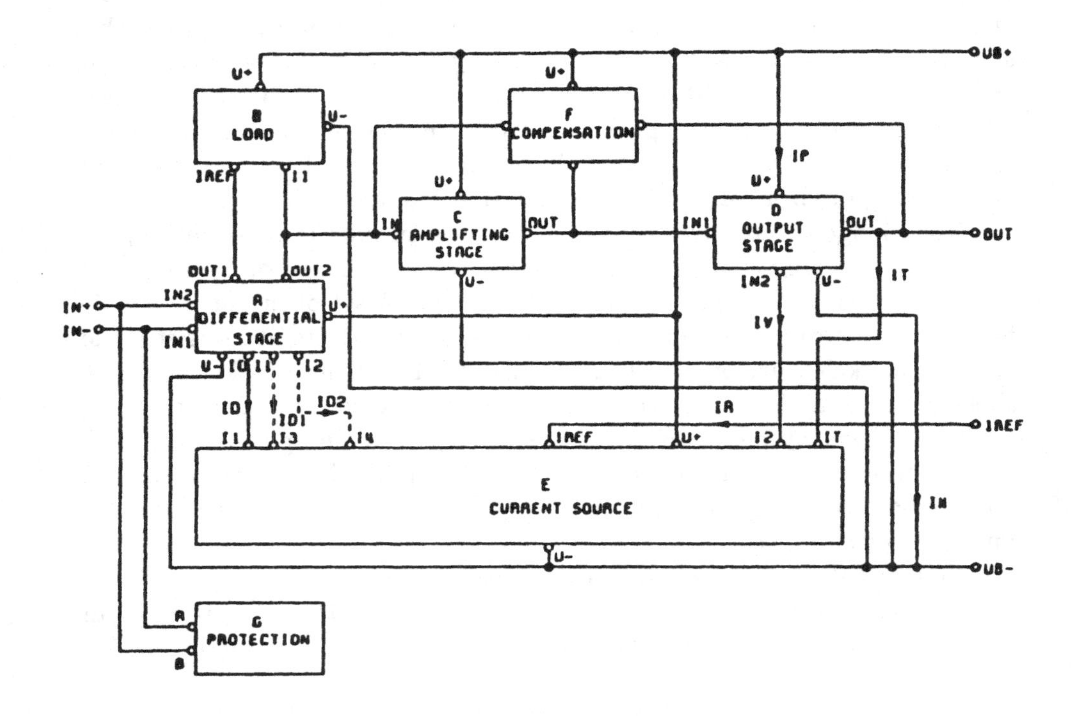


Figure 22: Skeleton representation of an opamp.

| Objective | Condition | Action |
|---|---|---|
| Input current | IF ( Ic > 0 ) | THEN default ELSE pnp-type stage |
| Offset current | IF ( Ios < Ios_val ) | THEN Darlington stage ELSE default |
| Input noise current | IF ( In < In_val ) | THEN npn-type Darlington input stage ELSE default |

Table 4: Input stage selection rules.

| Objective | Condition | Action |
| --- | --- | --- |
| Input stage | IF ( idev_type := pnp ) | THEN npn-type active load ELSE pnp-type active load |
| Offset voltage | IF ( Vos < Vos_val ) | THEN Wilson current mirror ELSE default |
| Slew rate | IF ( Sr > Sr_val ) | THEN resistive load ELSE default |

Table 5: Load stage selection rules.

| Objective | Spec. value |
| --- | --- |
| Input voltage swing | -8 V / +8 V |
| Output voltage swing | -8 V / +8 V |
| Offset voltage | < 2.2 mV |
| Offset current | < 100 nA |
| Supply current | < 1 mA |
| Input current | < 0.3 nA |
| Input resistance | - |
| CMRR | - |
| PSRR | - |
| Open loop gain | > 90 dB |
| Transit frequency | > 150 kHz |
| Distortion | min. |
| Input noise current | < 4 pA $\sqrt{Hz}$ |
| Slew rate | > 0.1 V/us |

Table 6: Design objectives for opamp assembly.

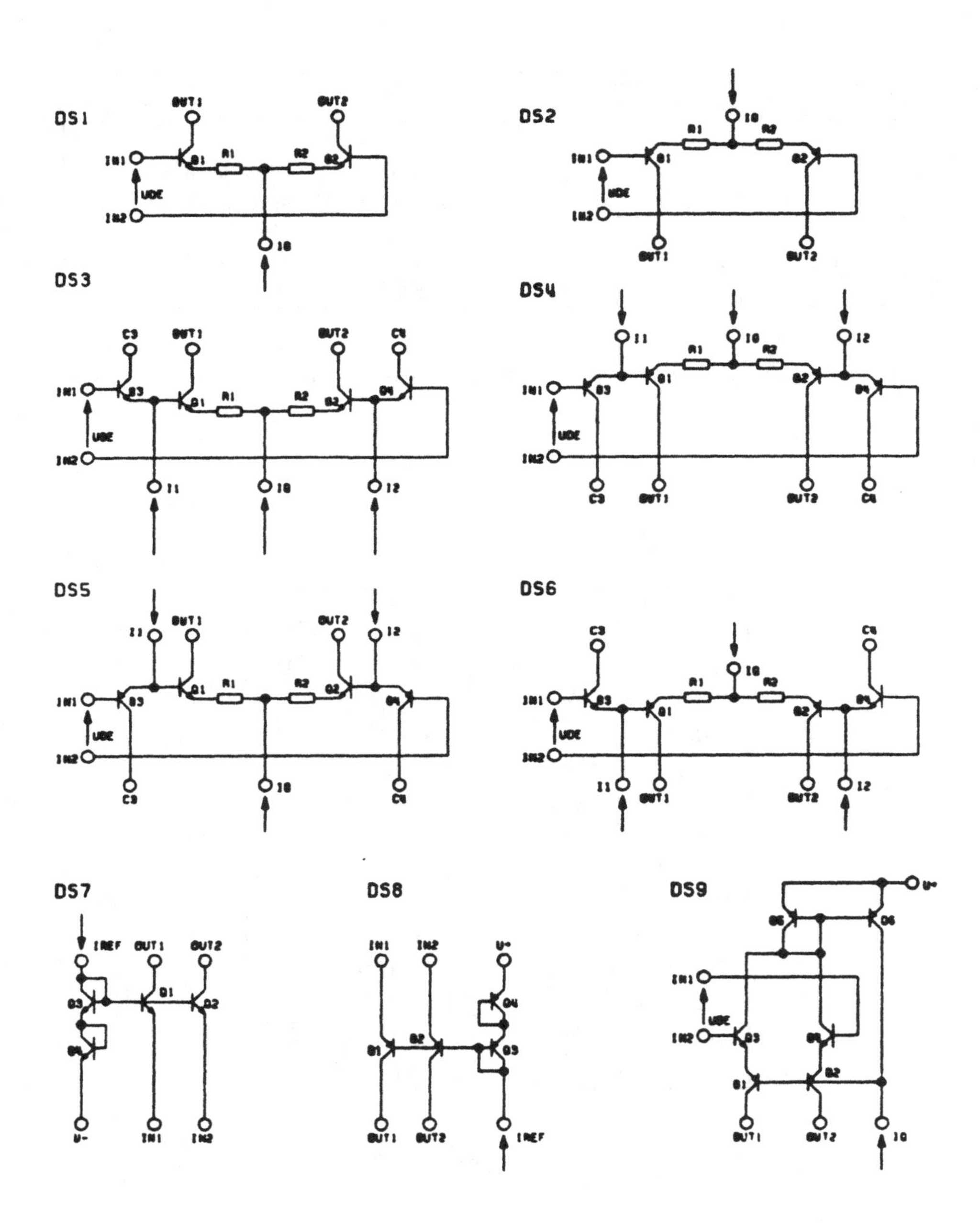


Figure 23: Set of input stage implementations.

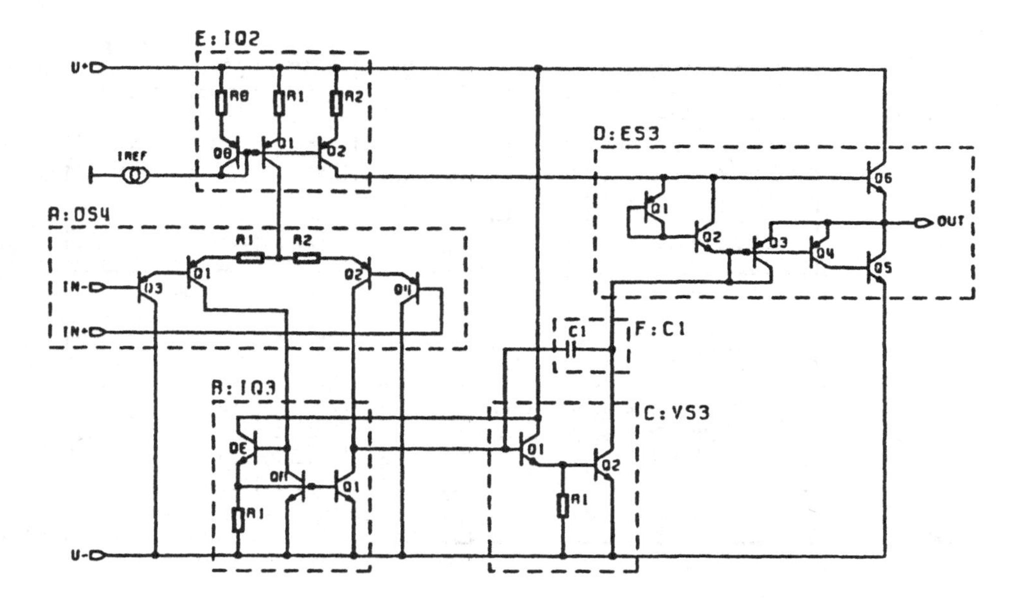


Figure 24: Schematic of the assembled opamp.

of the default implementation labeled DS1.

The set of objectives for opamp cells and the specification values for a design application are summarized in Table 6. The opamp implementation resulting from theses specs is depicted in Fig. 24. At this point the assembly of the cell is completed with respect to circuit topology. The values of components within the chosen skeleton implementations (i.e. resistor and capacitor values) are now to be established.

This task is best done by means of the nominal design tool presented in the electrical design section of this chapter. The final simulation of the assembled cell proved that all specifications were met. The frequency domain analysis result for the assembled opamp is summarized in the Bode diagram of Fig. 25.

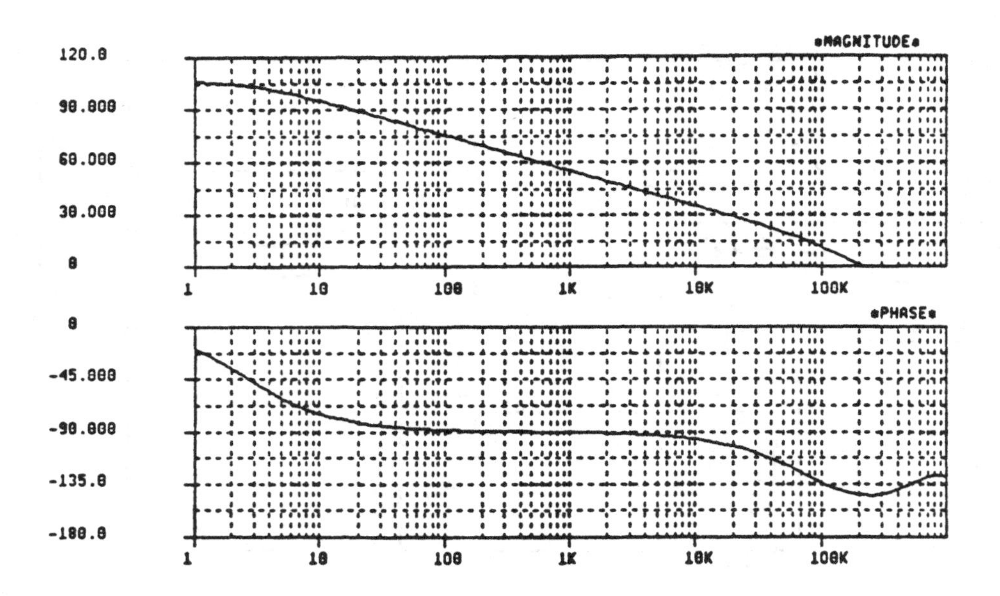


Figure 25: Frequency domain response of the assembled opamp.

# References

[1] A. Stuermer, "The measurement and evaluation system MAUS and its application to Spice model parameter calculation" (in German). In "2. Symposium Simulationstechnik", F. Breitenecker and W. Kleinert (Eds.), Springer-Verlag, Berlin-Heidelberg, 1984.

[2] G. D. Hachtel, T. R. Scott, R.P. Zug, "An interactive linear programming approach to model parameter fitting and worst case circuit design". IEEE Trans. on Circuits and Systems, Vol. CAS-27, pp 871-881, 1980.

[3] B. Nye, A. Sangiovanni-Vincentelli, J. Spoto, A. Tits, "DELIGHT.SPICE: An optimization based system for the design of integrated circuits". Proc. IEEE Custom Integr. Circuits Conf., pp. 233-238, 1983.

[4] K. J. Antreich, S. A. Huss, "An interactive optimization technique for the nominal design of integrated circuits". IEEE Trans. on Circuits and Systems, Vol. CAS-31, pp. 203-212, 1984.

[5] S. A. Huss, "An interactive, optimization based tool for the nominal design of integrated circuits". Proc. 11th Europ. Solid-State Circuits Conf., pp. 209-213, 1985.

[6] H. Kling, "Breitbandverstaerker". German Patent No. P 3840854.

[7] J. Burkhart, J. Schomers, "A monolithic bipolar 16x16(+16) crosspoint matrix with optimized power consumption". Proc. 14th Europ. Solid-State Circuits Conf., pp. 252-255, 1988.

[8] M. Gerbershagen, S. Huss, H. Ruff, "Maniac: A program for performance evaluation of analog circuits" (in German). ITG-Fachberichte, Vol. 98, pp. 249-252, 1987.

[9] I. Rugen, C. Schroeck-Pauli, M. Gerbershagen, "An interactive layout design system with real-time logical verification and extraction of layout parasitics". IEEE J. Solid-State Circuits, Vol. 23, pp. 698-704, 1988.

[10] J. Gruber et al., "Monolithic integrated circuits for 565 Mbit/s and above". Proc. 10th Europ. Solid-States Circuits Conf., pp. 79-82, 1984.

[11] K. M. Just, J. M. Kleinhans, F. M. Johannes, "On the relative placement and the transportation problem for standard cell layouts". Proc. 23rd Design Autom. Conf., pp. 308-313, 1986.

[12] M. Gerbershagen, S. Huss, M. Laage, I. Rugen, "A hierarchical cell based engineering system for the design of semi-custom analog integrated circuits". Proc. IEEE Custom Integr. Circuits Conf., pp. 2.2.1-2.2.4, 1988.

[13] W. Borutzky et al., "A novel approach to CAD of analog cells". Proc. IEEE Int. Symp. on Circuits and Systems, pp. 1791-1794, 1988.

## ACKNOWLEDGMENTS

The work described in this chapter was performed during the past five years by the technical staff of the AEG Design Center. I am pleased to acknowledge the help from my colleagues in preparing this paper.

# CHAPTER 6

# SILICON COMPILER TECHNOLOGY FOR SC FILTERS

W. Martin Snelgrove
Department of Electrical Engineering
University of Toronto
Toronto, Ontario
Canada M5S1A4

## 1 INTRODUCTION

"SiComp" [1,2] was an early University of Toronto prototype of a synthesis program for switched-capacitor filters. It took a design from specifications all the way down to layout. Several similar packages now exist, some of commercial quality (e.g. SYNSCAP [3]).

Filter synthesis is a natural niche for analog CAD because it is a mathematically well-developed art (so that there is a complete body of "knowledge" available in ready-to-use mathematical or algorithmic form), and because it is a sufficiently arcane area of knowledge to motivate system designers to use a tool rather than "just" do the work by hand. These fortunate characteristics make filter synthesis something of a bellwether for analog CAD in general.

The overall package contains about 40 000 lines of code and was developed over a period of ten years by half-a-dozen different students. The bulk of this effort, though, was general-purpose filter research: the effort specific to s Switched-C layout synthesis was less than a man-year.

The package produced a fifth-order elliptic filter that worked on the first run: in this sense it was a success. On the other hand, it has not been used commercially as a whole (although most of its component parts have). Commercial acceptability of this type of package will take close integration with other CAD software and an ongoing effort to incorporate new filter and circuit techniques as rapidly as possible. No "static" package can hope to interest designers for very long.

Our experience with this project has contributed to the design of a new package - XFilter - that is improved in breadth of capability, software engineering, and human interface. This chapter describes both SiComp and XFilter.

# 2 OVERALL STRUCTURE

Filter design proceeds roughly as follows:

| | | | |
|---|---|---|---|
| *from* | system needs | *via* | manual design |
| *to* | filter specifications | *via* | approximation software |
| *to* | a transfer function | *via* | circuit synthesis |
| *to* | a netlist | *via* | layout synthesis |
| *to* | a layout. | | |

A few things need to be noticed about this sequence:

- it purports to be linear, without any of the long "snakes and ladders" paths commonly found in design.
- the first step is rarely automated.
- several distinctly different kinds of software are involved.

Each of these points is worth consideration before plunging into a more detailed look at one implementation of the individual steps.

First, the real design process is not entirely loop-free, and there are some backward dependencies to consider. Certain types of circuits can only implement a subset of possible transfer functions, and system designers may decide to restructure the overall system when they see a proposed filter layout or look closely at the transfer function. The existence of these backward loops can lead the design of a CAD package in two different directions: either to make the forward path fast so that an engineer can afford to loop many times, or to automate the loop itself. The first option leads one to concentrate on efficient algorithms and to structure a package as a design exploration system; the main representative of the second type of thinking is provided by the circuit-level optimizer, which is sometimes used to unify the approximation and circuit synthesis steps.

Second, the fact that specifications are manually prepared suggests the existence of a gaping hole in CAD coverage. Because applications are many and varied no single approach will suffice, but filters are used in a finite number of ways. Some consideration given to each major application (especially anti-aliasing and smoothing) could help to fill the gap. The reason to do this is to help a system designer who does not know filters to frame specifications properly.

Last, the variety of types of software involved affects project management: it is easy enough to partition the programming effort somewhat by type, but interfaces have to be rigorously controlled to avoid showing the seams.

Our response to the backward dependencies has been to promote design exploration; for the "CAD gap" we have a few preliminary ideas for integrating filter synthesis more closely into general-purpose system design software; and lastly, our early efforts used the variety of types of software to give us the flexibility to do research on each subject independently, but we are now integrating the components more closely. Now, onto the components themselves.

# 3 APPROXIMATION

It is quite common in simple filter packages to provide only classical approximation functions (Butterworth, Chebyshev, elliptic and so on). This is very straightforward, involving only a few days of software effort, but not really adequate for modern needs.

We have three generations of approximation software: "REMEZ", "xfer", and "CoCo". The first implements minimax approximation for filters with arbitrary stopbands and equiripple or maximally flat passbands. "xfer" is more general but slower: it allows simultaneous least-$p^{th}$ optimization of transfer functions subject to quite arbitrary specifications on amplitude, group delay and time response. The third version, "CoCo", primarily cleans up the software engineering side of xfer. It features an easily extended command language generated by a compiler-compiler.

SiComp used REMEZ, and XFilter is being built on CoCo. The extra flexibility of the newer software is particularly useful for modem and sampling applications, while REMEZ works well for more traditional audio filtering. In particular, older approximators handle only flat passbands, which are appropriate when filters are used for frequency-division multiplexing. For anti-aliasing work, though, compensation of sample-and-hold errors ($\frac{sin\omega T}{\omega T}$ distortion) calls for more complicated passbands; instrumentation and pulse applications call for simultaneous specifications on gain and phase; and data communications applications need to specify time and frequency responses simultaneously.

REMEZ uses a traditional dialogue to elicit information from the user, which is augmented by a simple (and compatible) command language which can be used to override the usual sequence of questions. Thus REMEZ usually proceeds by asking for passband specifications, then for stopband, then suggested order, and so on. At any time, however, a designer can interrupt the sequence and change an earlier response or ask for a classical approximation.

This mixture was an attempt to balance the needs of novice users (who need the guidance offered by the question-and-answer format) with those of expert users (who want the flexibility that comes from a command language). This range of users is an important issue in CAD: analog CAD packages tend to have a small market in the first place, so getting them used calls for broadening their appeal by any means available; but even more importantly, expert filter designers start as novices and should be helped along in their development by the tools they use. A spinoff advantage of catering to a wide range of skills is that any given engineer may do filter design only occasionally, and has to be welcomed back even with rusty skills.

XFilter deals with this high "user dynamic range" by having a menu/ graphics front end for novice users and conventional problems, which drives a command-based applications package that can be used directly for more difficult problems. Figure 1 is a screen-dump of a filter design session using XFilter: it shows a command window headed by a menu bar, together with a graphics window showing the current specification and approximation. Note that this scheme not only gives a user a simple graphical interface for easy problems together with a powerful one for tougher designs, but (by translating the graphical "language" to the textual one as the designer watches) teaches the syntax and praxis of the powerful technique in a natural way, based on the intuitive graphical scheme.

# 4 EVALUATING APPROXIMATIONS

Any kind of non-routine design requires that the designer evaluate some function derived from the transfer function T(s) produced by the approximator. It may be necessary, for instance, to look at the effect of

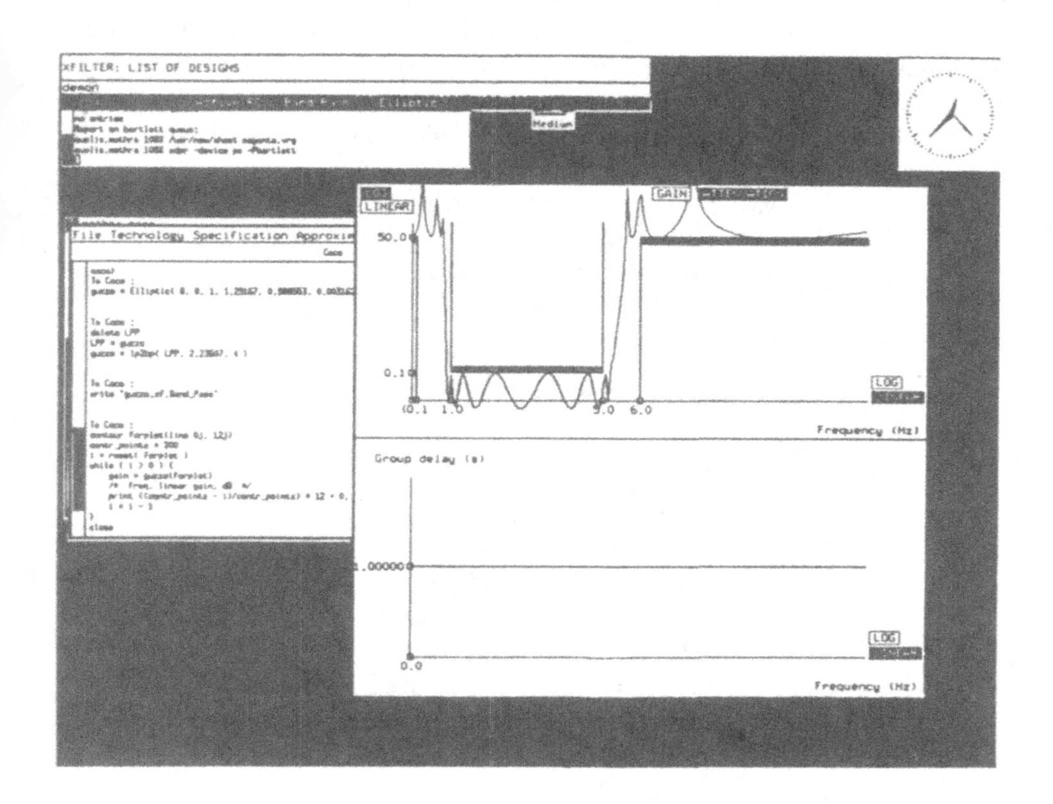


Figure 1: An XFilter Screen

changes in pole or zero positions, or to evaluate the overall response of the filter in cascade with a channel, or to compute some figure of merit. These things are necessary just because design is not as linear as the "overall structure" of filter design suggests.

Evaluation is done both in SiComp and XFilter by providing "programmable calculators". SiComp uses reverse Polish notation while XFilter is a little like C. Figures 2a) and 2b) show typical expressions for each calculator. In the long run, filter synthesis has to be seamlessly integrated into a much more general design exploration package, so that evaluating the effect of a filter decision on the overall system is straightforward.

```
eval i3 s * 0.213 * v2 + store 1
reg 1 real result=realpart
reg 1 imag result=imagpart
s s * -1 * h( db result=h2
type print end
```

Figure 2(a): The RPN calculator in SiComp

```
proc error(){
        i = 0
        while ( i < $1 ) {
        point = i / $1 * j
        desired = inv_sinc( point )
        l2mag = mag( l2( point ))
        l10mag = mag( l10( point ))
        error2 = ( l2mag - desired ) / desired * 100
        error10 = ( l10mag - desired ) / desired * 100
        print desired, " ", error2, " ", error10
        i = i + 1
    }
}
```

Figure 2(b): Calculator function in XFilter

An example of a data communications specification that the current technology does not even evaluate well (let alone synthesize) is "eye opening", a measure of time response to random pulse streams. An even tougher specification to handle would be bit-error rate, which is affected by subtle interactions between the filter and its surrounding circuitry.

# 5 NETLIST (CIRCUIT) SYNTHESIS

There are infinitely many circuits, in any given technology, realizing any given transfer function. Some of these circuits are better than others as to component cost, signal/noise performance, manufacturing yield, tunability and so on. The two general approaches to choosing among circuits are based on systematic synthesis and optimization respectively.

We take the synthesis approach in both SiComp and XFilter. Synthesis for SC filters can be thought of at two levels: a structural level at which filters are decomposed into primitive blocks like integrators and summers; and a circuit level at which these blocks are further decomposed into amplifiers, switches, and capacitors. At each level there are half-a-dozen popular choices with different strengths and weaknesses, and a good tool should provide a wide choice (for the expert) together with reasonable defaults (for the novice). SiComp handles only ladder simulation at the structural level, and produces netlists for three different circuit structures.

Of these two, the structural level is more difficult to code well than the circuit level: the techniques are mathematically more sophisticated and numerically less stable. On the other hand the circuit level is more subject to technological change and therefore liable to frequent revision. In SiComp the bottom-level circuit synthesis code was written by a circuit designer, not by a programmer. As long as this can be done with the technology-sensitive parts of the code, filter synthesis software can be kept up to date. It is usually practical because circuit researchers typically reduce this phase to a matter of plugging numbers into closed-form formulae.

It is also possible to produce netlists by circuit optimization. This has the advantage that almost arbitrary (and realistic) cost functions can be minimized, so that truly optimum designs can be produced; it has the disadvantages that the search space is in general of a very high dimension, and that practical problems accidentally omitted from the cost function will get "traded away".

The synthesis and optimization approaches may in general be complementary, with synthesis providing a good starting point and optimization refining it.

# 6 LADDER DESIGN

The "structural level" synthesis can be done in several ways, but some of the best (as to noise and sensitivity performance) filters are derived from passive LC ladders. This means that a transfer function is first synthesized as an LC netlist, then transformed to an equivalent structure, which is then implemented in Switched-C technology.

There are two issues of interest here: LC synthesis itself and the fact that designers work by transforming circuits.

LC synthesis is just the kind of thing that system designers dislike about filter theory: it is old, mathematically tricky, only works for certain types of transfer functions, and only worth a few dB of performance. That means that we should automate it thoroughly or forget it completely, rather than trying to explain the significance of "a partial removal at infinity" to someone with a circuit to build. The hardest thing about automation here is that there are usually a few choices to make, and the theory for how to make them involves copious arm-waving. The best thing to do would probably be to encourage manual exploration by maintaining a prioritized list of choices on the designer's behalf.

The fact that much design knowledge is in the form of circuit transformations is methodologically interesting. One can view these just as "transformation rules", like productions in a grammar, or look at them as a peculiar use of abstraction that inverts the concept of a design hierarchy. The rules used to translate LC ladders to SC circuits take a specific circuit, abstract it to a set of defining equations, then find a different implementation of the equations. Viewed in terms of hierarchy, this is a process by which we move from a particular "leaf cell" up to a parent, then down again to a sister leaf cell. At the least, this means that designers like to explore a tree of possibilities by moving between the leaves (rather than doing top-down design, or successive refinement). More than that, though, this common practice challenges the concept of a design hierarchy, because different types of transformations induce their "parents" in different ways– so that each cell can have many parents. Design has the structure of a general graph, not of a tree.

# 7 NETLIST EVALUATION

Just as a general-purpose calculator program is need to evaluate and edit transfer functions, so a circuit simulator is needed to "experiment" with synthesized netlists. SiComp emits a SWITCAP [3] format circuit description. The main problem here is that there is a plethora of competing netlist standards. Again, proper integration into a much more general design environment is the ideal to strive for.

# 8 LAYOUT SYNTHESIS

We produced a layout based on parameterized standard cells. The circuit was decomposed into blocks (one block was an op-amp, another an integrating capacitor, and the rest were standard combinations of switches and input capacitors) which were designed so that sensitive nets would butt, while amplifier outputs were routed by a simple standard-cell router intended for digital work. Figure 3 is a schematic that shows the overall floor plan of a typical block.

This is certainly not the densest way to do a layout, but then standard cell designs in digital technology seem to have been quite successful despite taking an area (and performance) penalty. The advantages that come from using proven cells, from allowing designers to work on cells independently of each other, and from cell re-use apply as much to analog as to digital technology. If the objective is to get the design out the door quickly, or to work in a new technology quickly, then an extra $0.1mm^2$ may not be a big problem. Chips that then go to high production volumes may call for tighter layout, but in these circumstances the cost and speed advantages that CAD can offer are less important than in low-volume cases, while manual layout has the advantage of being up-to-date with the detailed requirements of the technology and the chip.

At the moment we are concentrating on breadth (by adding digital and continuous-time capabilities up to the netlist level) rather than doing more work on layout. A good "analog sticks" abstraction would be an attractive way to get back into layout work.

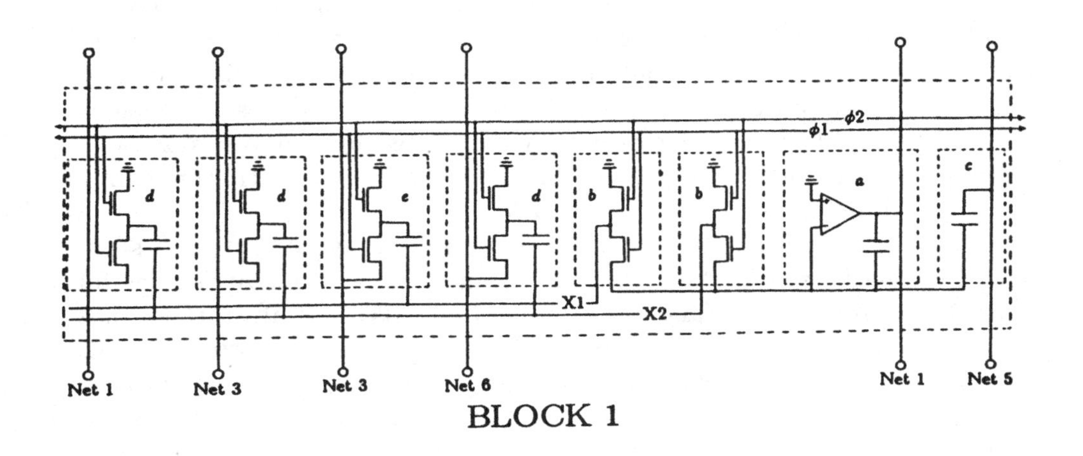


Figure 3: Floor Plan of a SiComp layout block

# 9 CONCLUSIONS

SiComp was an interesting experiment in analog CAD, but strictly an experiment. The software and electronics techniques involved are fairly straightforward, but making a real tool involves a wide range of issues. Most of the hard problems have to do with the fact that filter design (like any particular type of analog design) is just a small part of system design. From this comes the need for close integration with other CAD software and the need for an intuitive human interface. The technology of filter synthesis has been proven practical, but the "business" side - making something that will be widely enough used to justify the effort of writing it – needs work.
Analog synthesis is trickier than it looks.

## References

[1] G.V. Eaton, D.G. Nairn, W.M. Snelgrove. A.S. Sedra, "SiComp: A Silicon Compiler for Switched-Capacitor Filters", ISCAS-87, Philadelphia, May 1987, pp. 321-324

[2] G.V. Eaton, "SiComp: A Silicon Compiler for Switched-Capacitor Filters", MASc thesis, University of Toronto, 1986.

[3] D.K. Bui, T. Mishawa, "Application of SYNSCAP in designing switched- capacitor filters", Bell Labs Technical Memorandum 52173-880731-01-TM, July 1988

# CHAPTER 7

# CAD-COMPATIBLE ANALOG SYSTEM DESIGN: A NEW DESIGN CONCEPT

Mohammed Ismail
Solid-State Microelectronics Laboratory
Department of Electrical Engineering
The Ohio State University
Columbus, OH 43210-1272
U.S.A.

## 1 INTRODUCTION

This chapter provides an entirely new concept of analog design. It advocates *analog design for CAD/automation* or *CAD-Compatible analog design* in a way that will bridge the existing gap between circuit design- and CAD-oriented work in the analog field, at both the circuit and layout levels, and bring analog VLSI design a step closer towards automation. It also unveils strong and interesting connections between programmability, computer-aided design and physical implementation of analog VLSI circuits. This leads to analog system designs that rely on the interconnection of simple reconfigurable programmable *fixed* analog cells instead of *parameterized* analog cells. This new design concept gives an important dimension to work discussed in previous chapters. For example, using the new concept, the work on automatic synthesis of fixed cells such as op-amps, can easily be extended to automatize analog design at the system level. Design examples of continuous-time analog MOS integrated systems are given to demonstrate the new design concept.

# 1 ANALOG DESIGN FOR CAD

At the layout level, parameterization of an analog cell to meet a different set of specifications require a full-custom re-design of the layout, e.g. re-size a capacitor to achieve a different time-constant. Parameterization, as such, represents a major hurdle if one is to automate the design process of analog integrated circuits. An alternative approach to this is to assign a fixed area in the layout to accommodate a circuit component for different specifications (component sizes). Obviously, this is more compatible to semi-custom or automated CAD design, but it may result in large waste in silicon area [1,2]. To solve this problem, sophisticated layout compaction tools, that are primarily digitally oriented, may be used to minimize area waste resulting in an area-efficient layout. Although digital placement and routing tools are to some extent acceptable for analog circuits, digital layout compaction tools may result in serious analog performance degradations due to potential noise, crosstalk and supply rejection problems associated with this process.

A completely different approach is to investigate programmability of analog circuits in relation to CAD and physical implementation. It is interesting to unveil a strong connection between programmability and "CAD-Compatible" analog circuit design that seems to be overlooked in analog design work so far. The programmability we refer to here provides means to parameterize an analog cell without having to alter its physical layout, i.e. hardware or software means to access the layout on- or off-chip. This way, one may be able to come up with a nominal fixed layout where parameterization of the analog cell to meet different specifications is achieved via programmability, instead of full- or semi-custom redesign of the layout. In other words, analog design as such will rely on the interconnection of simple "programmable" fixed analog cells instead of "parameterized" analog cells. This will bring us an entirely new view of analog design; viz., *analog design for CAD/automation or CAD-compatible analog design* in a way that will bridge the existing gap between design- and CAD-oriented work in the analog field, at both the circuit and layout levels.

This new design concept gives an important dimension to research work in the area of automatic synthesis of analog integrated circuits. A major part of this work [3] has been focusing on the automatic synthesis of fixed cells such as op-amps, e.g. OPASYN [4]and OASYS [5]. Using this new concept, the extension of automation ideas to the system level is easily achievable since the whole system will enjoy a fixed physical layout.

## 2 INHERENT PROGRAMMABILITY IN CONTINUOUS-TIME ANALOG INTEGRATED CIRCUITS

It is important to note that programmable analog integrated circuits using discrete-time MOS Switched-Capacitor (SC) techniques have been reported [6]. The motivation behind these SC programmable structures was merely to provide convenient means to tune the circuit and did not unveil the interesting connections brought about in the previous section between programmability and analog design for CAD. Since the design of continuous-time analog MOS integrated circuits is largely motivated by problems encountered in SC circuits due to sampling, it is interesting to briefly discuss programmability in SC networks in relation to their physical implementation. The clock frequency in SC circuits provides a very convenient means to program frequency specifications, e.g. cut-off frequency of a SC filter. However, when it comes to programming of other design parameters such as DC gain, pole-Q, etc., capacitor arrays are used which provide at best *discrete* programmability/tuning at only a set of prescribed values of these design parameters [6]. The range of programmability is determined by how large these capacitor arrays are. As a result the use of such arrays results in larger chip areas where only a portion of the array is used to meet a certain prescribed value of the design parameter.

Continuous-time analog MOS integrated circuits, on the other hand, use MOS transistors as *voltage-controlled resistors* [7-19] in such a way that the nonlinearities in the MOS drain current equation are partially or totally canceled. It will be shown in the following sections that continuous-time analog MOS integrated circuits are inherently programmable. Specifically, it will be shown that as a result of the fact that "MOS resistors" are voltage controlled, all design parameters of an analog circuit can indeed be conveniently programmed over a *continuous* range. The continuous-time MOS analog cells of Fig. 1 can be looked at as fixed analog cells in which design parameterization is achieved, not by resizing circuit components, but by changing the DC control voltages at the MOSFETs gates.

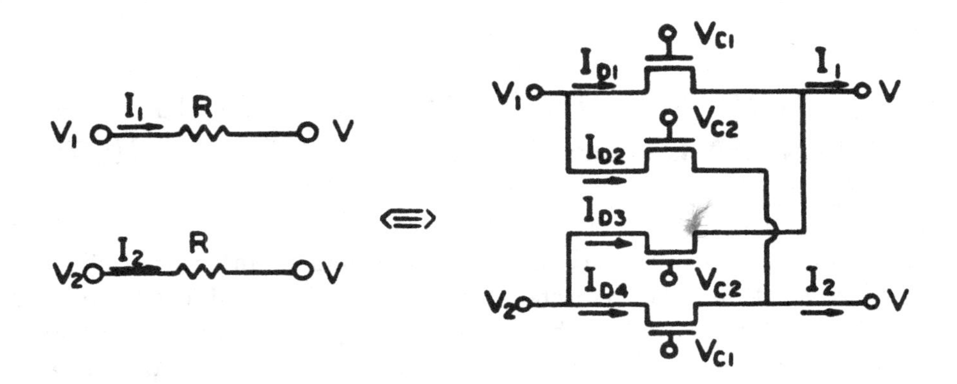


Figure 1(a): The Double-MOSFET method [7,11] for nonlinearity cancellation. The 4-matched transistors produce a linear current difference $I_1 - I_2$ where all even and odd nonlinear terms are cancelled and simulate a resistor pair $R$ with $R = 1/\mu C_{ox}(W/L)(V_{C1} - V_{C2})$.

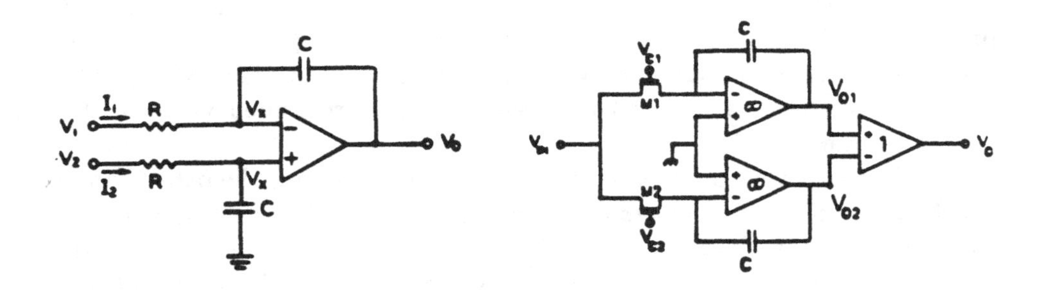


Figure 1(b): A differential integrator cell where the input resistor pair, $R$, is replaced by MOS transistors according to Figure 1(a). Shown to the right is another integrator cell obtained using a MOS-domain design methodology [12]. It achieves complete cancellation of nonlinearities by using only 2 MOS transistors. It also has good supply rejection and is less sensitive to device matching.

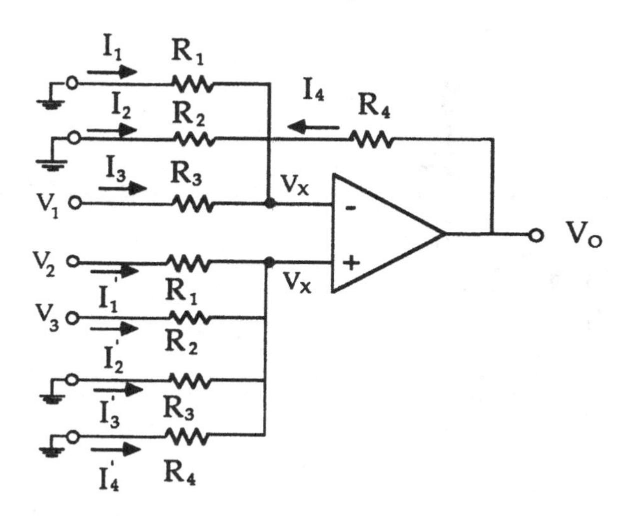


Figure 1(c): A weighted summer block. Resistors appear in matched pairs which are replaced by MOS transistors according to Figure 1(a). Each pair uses 4-matched transistors with specific (W/L) ratio for the pair.

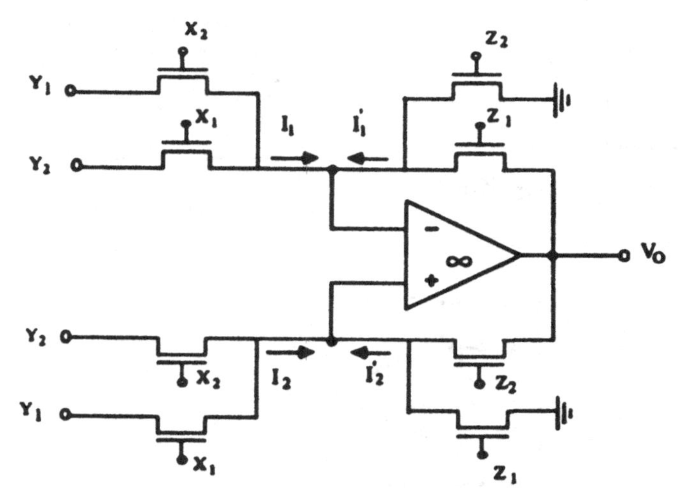


Figure 1(d): A MOS multiplier/divider cell with $V_O = \left[\frac{(W/L)_i}{(W/L)_o}\right] \frac{\Delta Y \Delta X}{\Delta Z}$ where $(W/L)_i$ and $(W/L)_o$ are the aspect ratios of the 4-matched transistor cells at the input and output respectively. $\Delta Y = Y1 - Y2, \Delta X = X1 - X2$ and $\Delta Z = Z1 - Z2$ are differential inputs which can also be used as grounded voltages when $Y2 = X2 = Z2 = 0$ with proper DC biasing [18].

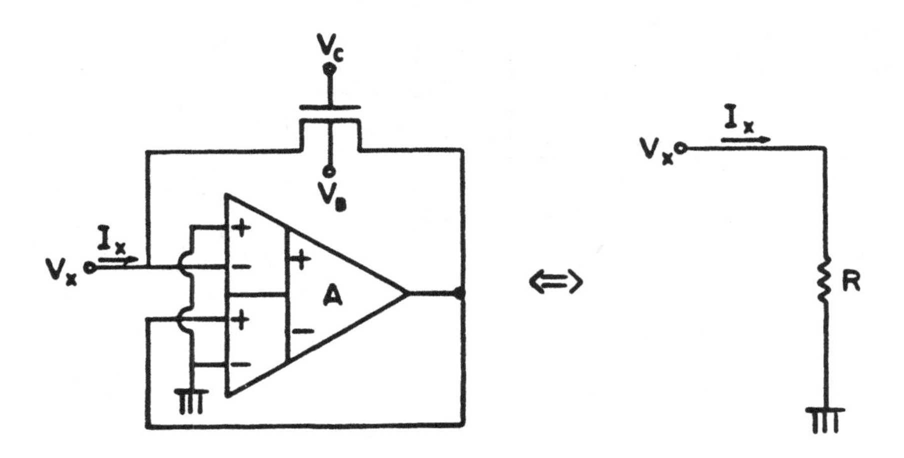


Figure 1(e): A new DDA-based MOS grounded resistor cell. It uses only 1 MOSFET and requires no matching for nonlinearity cancellation.

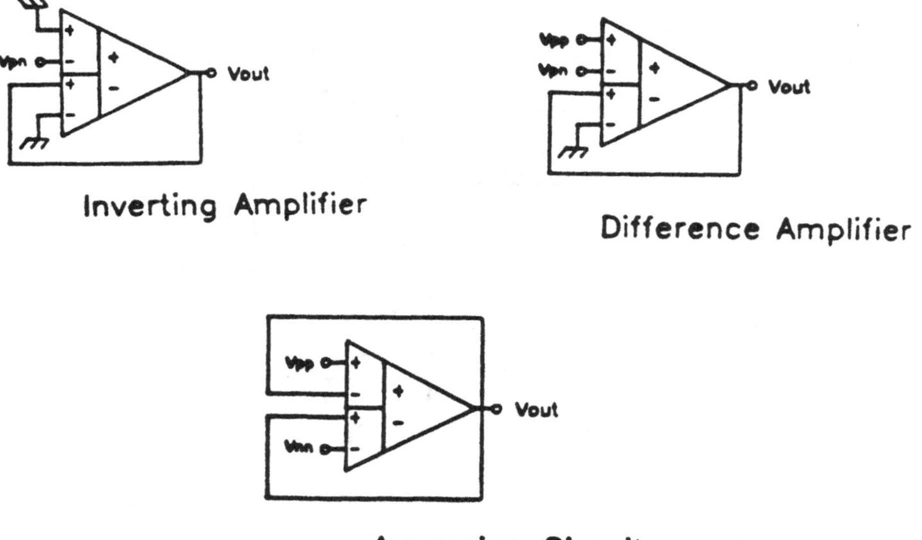


Figure 1(f): Three simple DDA-based fixed cells [20]. The difference amplifier is used to implement the unity gain differential amplifier in Figure 1(b). The inverting amplifier is used in the MOS resistor cell of Figure 1(e).

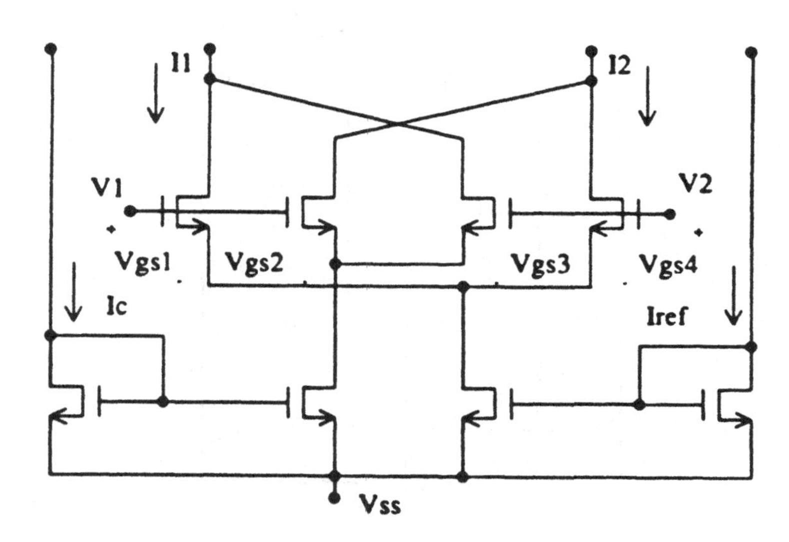


Figure 1(g): A difference input voltage, current controlled transconductor circuit for which:
$G = \frac{I_1 - I_2}{V_1 - V_2} = \left(\frac{\mu C_{ox} W}{2L}\right)(V_{GS1} + V_{GS2} + V_{GS3} + V_{GS4} - 4V_{To}) = 2KV$
where $V_{GS2}$ and $V_{GS3}$ are controlled by $I_C$. $G$ is constant if $V$ remained a positive constant over a particluar range of the difference input voltage [24].

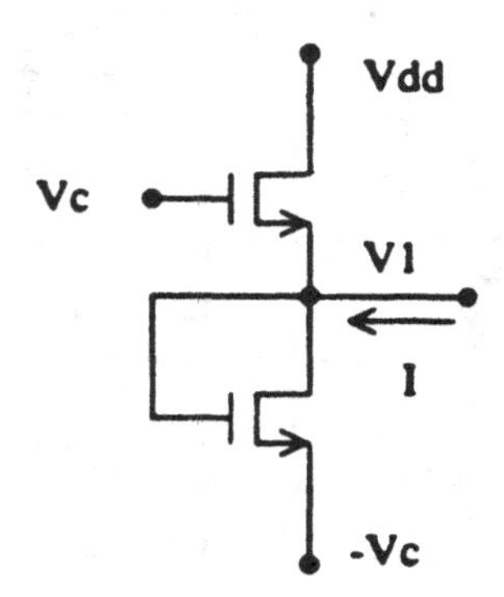


Figure 1(h): A grounded voltage-controlled linear resistor. Transistors $M_1$ and $M_2$ are operating in the saturation region. The value of this resistor, $R$, is defined by: $R = \frac{V_1}{I} = 1/\mu C_{ox}\frac{W}{2L}(V_C - V_{To})$ [24].

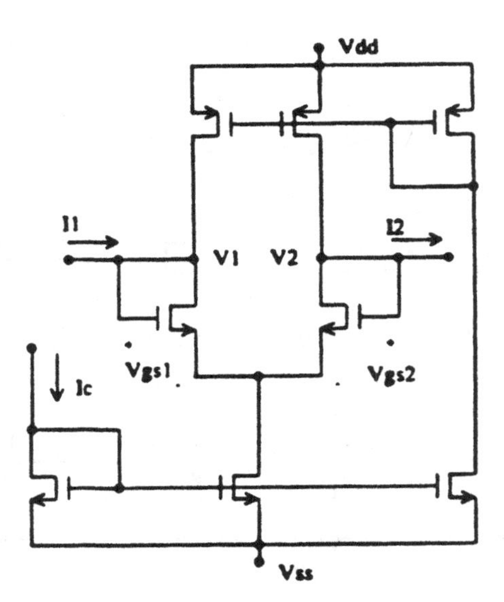


Figure 1(i): A simple current-controlled floating linear resistor. For the specific range of operation $R$ is given by: $R = \frac{V_1 - V_2}{I_1} = \frac{V_1 - V_2}{I_2} = 2L/\mu C_{ox}W(V_{GS1} + V_{GS2} - 2V_{To})$, and $V_{GS1} + V_{GS2}$ remains constant and is controlled by $I_C$ [24].

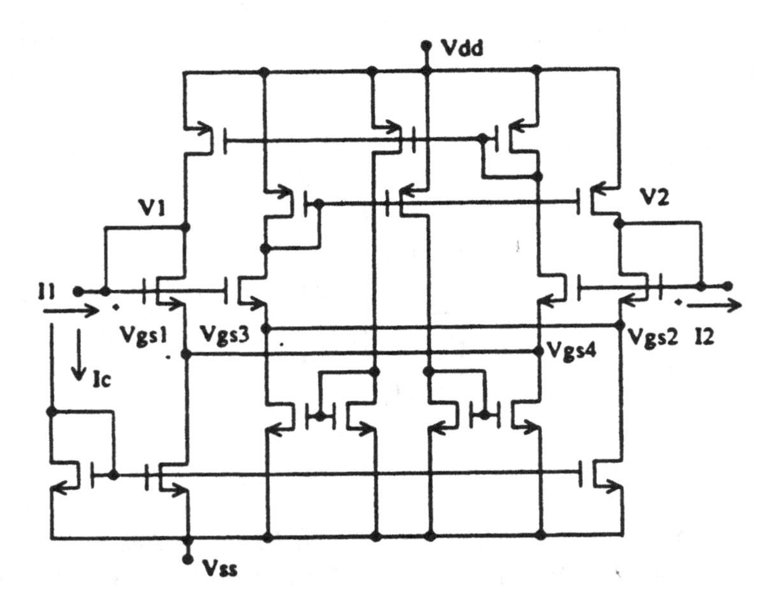


Figure 1(j): An improved realization of the resistor of Figure 1(j) [24].

In the following sections we will show that interconnections of some of these cells result in system-level designs where programmability of all design parameters is possible. It will also be seen that inherent programmability of continuous-time MOSFET-C [7] circuits together with their versatility, reconfigurability and modularity provide an example and set a new trend in analog integrated circuits where several design goals are simultaneously achieved to provide subtle circuit design that is both CAD-compatible and area efficient.

# 3 PROGRAMMABLE-Q MOSFET-C FILTER STRUCTURES

MOSFET-C continuous-time integrated filters [7-13] make use of fixed analog cells, such as the operational amplifier (op-amp) or the differential difference amplifier (DDA), in their design. The result is analog MOS architectures that are easy to design using available VLSI CAD tools and hence, require less design time with a high degree of success on first silicon. We have been able to set a general framework with several straight-forward design methodologies to obtain a MOSFET-C filter structure from a classical active-RC prototype [7], or design the filter exclusively in the MOS-domain [12]. We were also able to unveil some of the interesting features of MOSFET-C structures such as versatility, reconfigruability and their high-degree of modularity [2]. The more reconfigurable/versatile the analog circuit is the more it becomes like a digital cell. We created a library of well-defined analog parameterized cells, such as those shown in Fig. 1, which can be interconnected to achieve different linear and/or nonlinear functions. We used these cells to develop both linear and nonlinear system-level designs such as an all MOS universal filter structure [7] and an AM-amplitude modulator [19].

Here we apply the ideas discussed in the previous sections to the design of a second-order programmable-Q MOSFET-C structure which belongs to the two-integrator state-variable class of filters represented by the block diagram of Fig. 2(a).

The pole-frequency, $\omega_0$, and pole-Q, $Q$, of the filter are given by:

$$\omega_0 = \frac{1}{RC} \qquad\qquad Q = \frac{R_2}{R} \tag{1}$$

From the block diagram of Fig. 2(a), an active-RC prototype circuit which comprises two integrators connected in a loop is developed and

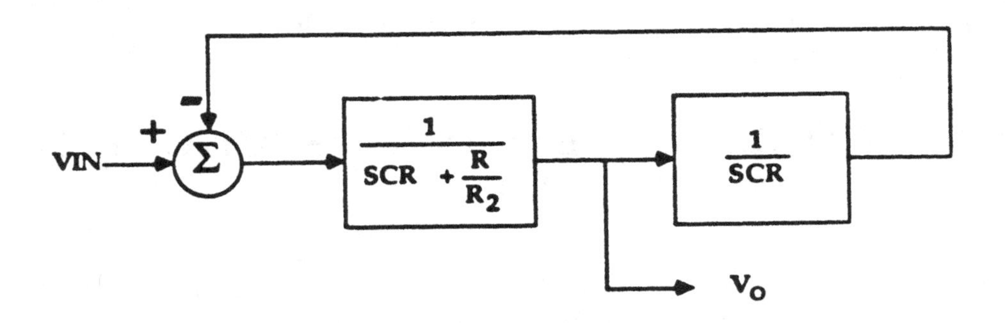


Figure 2(a): Block-diagram of a second-order biquadratic filter.

is then converted to an all MOS implementation by replacing resistors by MOS transistors, e.g. using the methodology of Fig. 1(a), where the equivalent resistance value is given by:

$$R_{eq} = \frac{1}{\mu C_{ox} \frac{W}{L} (V_{C1} - V_{C2})} \tag{2}$$

The above equations indicate that 4 MOS transistors, having an aspect ration $(W/L)$ to implement R, can be used to program $\omega_0$ for different design values using the control voltages $V_{C1}$ and $V_{C2}$ (refer to Fig. 1(a)).

The equations also indicate that independent programmability of $Q$ is impossible since it depends on R. This becomes clear if we use $V'_{C1}$ and $V'_{C2}$ to refer to the control voltages of the MOSFETs having an aspect ratio $(W/L)_2$ and used to implement $R_2$. Thus $Q$ of the MOS circuit takes the following form:

$$Q = \frac{\left(\frac{W}{L}\right)(V_{C1} - V_{C2})}{\left(\frac{W}{L}\right)_2 (V'_{C1} - V'_{C2})} \tag{3}$$

Now it is seen that $Q$ is a function of $V_{C1}$ and $V_{C2}$ which are used to program $\omega_0$. In fact, the only possible way to have a meaningful design for $Q$ is to use $V_{C1} = V'_{C1}$ and $V_{C2} = V'_{C2}$. This leaves us with the aspect ratio $(W/L)_2$ which must be altered if we are to realize a different design value for $Q$, since $(W/L)$ is used to realize a nominal design value for $\omega_0$. So, $\omega_0$ is easily programmed via $V_{C1}$ and $V_{C2}$ whereas every time we need to realize a different design value for $Q$, a redesign of the physical layout of the MOS transistors aspect ratio $(W/L)_2$ is mandatory.

In order to overcome this problem and allow for $Q$ programmability, we consider the modified block diagram of Fig. 2(b) in which a parameter $X$ is introduced. The design equations of the modified filter are:

$$\omega_0 = \frac{1}{RC}, \qquad Q = \frac{R_2}{RX} \tag{4}$$

and an active-RC implementation of it is developed as shown in Fig. 3 where a divider circuit is connected in the feedback loop of the two integrators [21].

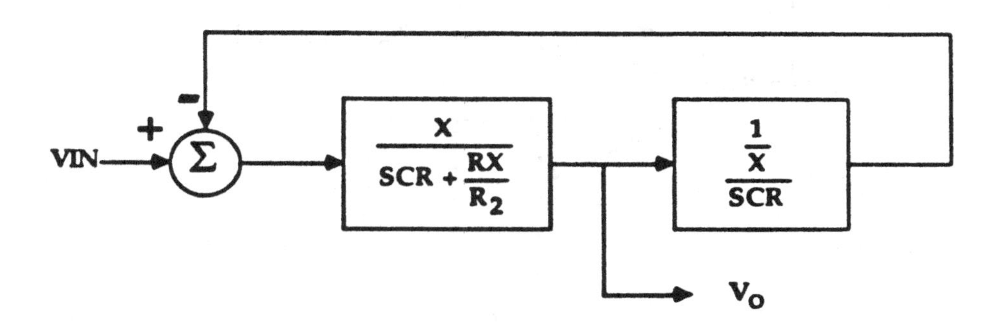


Figure 2(b): Modified diagram for Q-programmability.

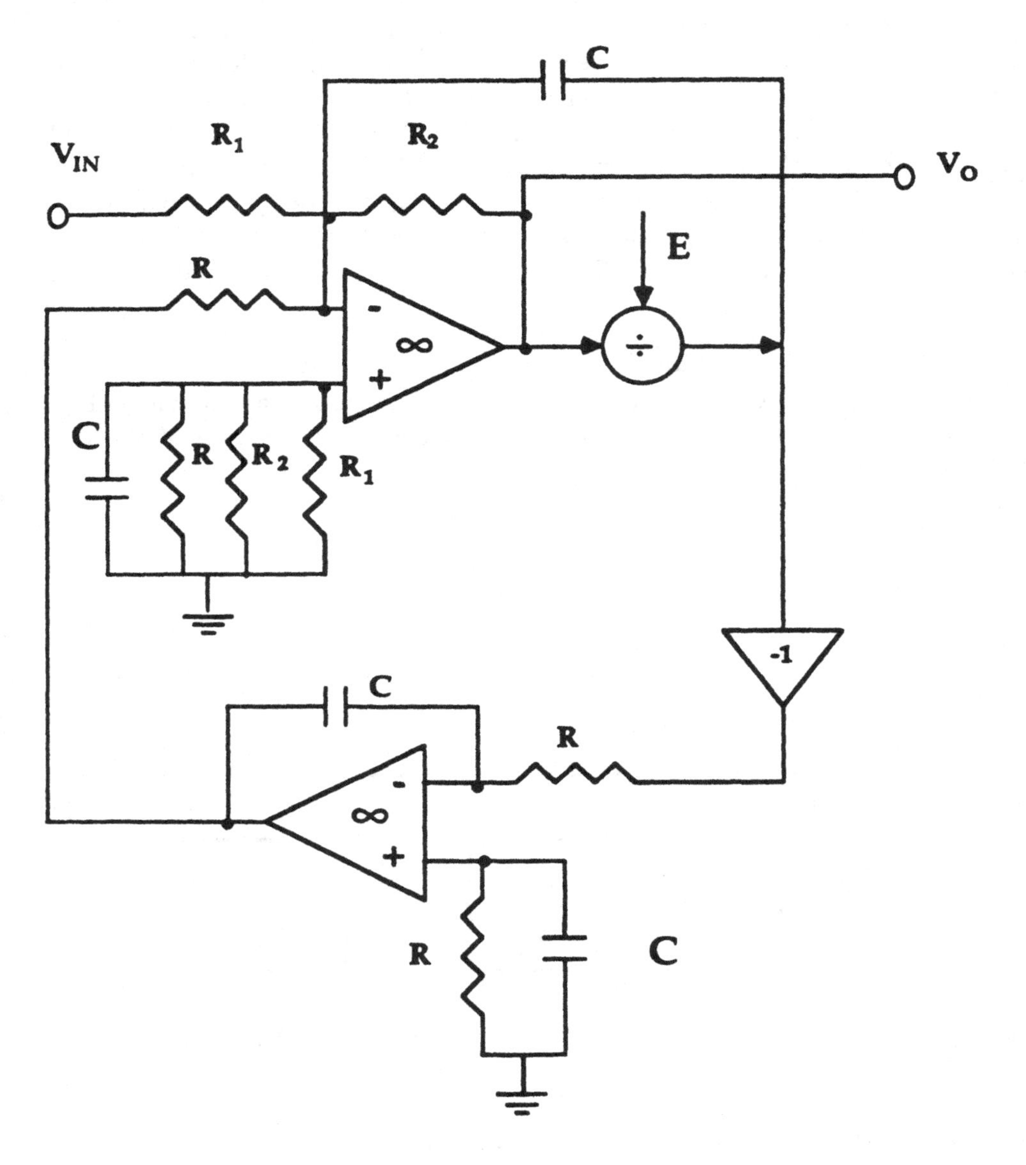


Figure 3: An active-RC prototype for a programmable-Q and $\omega_0$ filter structure.

A MOS implementation is obtained by replacing the matched resistor pairs of the filter by MOS transistors according to Fig. 1(a) and using the simple MOS multiplier/divider cell of Fig. 1(d). The new circuit allows versatile and independent programmability of both $\omega_0$ and $Q$. The expression for $Q$ becomes:

$$Q = \frac{(\frac{W}{L})(\frac{W}{L})_{in}(V_{C1} - V_{C2})_D}{(\frac{W}{L})_2(\frac{W}{L})_o E} \tag{5}$$

Where $(V_{C1} - V_{C2})_D$ are the DC control voltages of the divider circuit (the $X_1$ and $X_2$ voltages in the divider circuit), $(W/L)_{in}$ and $(W/L)_o$ are the aspect ratios of the divider input and output transistors, respectively, and $E$ is a DC signal and is an input to the divider which can be applied in a regular or a fully-balanced fashion. Both $E$ and $(V_{C1} - V_{C2})_D$ can be used to alter the design value of $Q$ and achieve a very wide range of programmability.

It is now clear that the whole filter can be designed using a fixed physical layout obtained to satisfy nominal design values for $\omega_0$ and $Q$ and that changing the design values is simply achieved by programming the respective control voltages in a very convenient manner. No need to alter the physical layout or to use layout compaction packages. The basic cells used in the implementation of the filter can be optimized, generated, placed and routed automatically resulting in an area efficient layout.

# 4 A PROGRAMMABLE MOSFET-C BUMP EQUALIZER

In the previous section, we had to modify the design of a MOSFET-C filter, by incorporating a divider cell, to achieve programmability. Here we show that in some applications, the MOSFET-C implementation is inherently programmable and interconnection of linear cells such as integrators and weighted summers is sufficient to achieve programmability.

An excellent example which we are currently investigating is the Bump amplitude equalizer. Bump equalizers, also known as Dip equalizers or variable amplitude equalizers, are widely used for amplitude equalization with applications in audio systems to equalize the entire audio spectrum of one sound system channel resulting in a proper acoustical response for the listener's ear [6]. Equalizers are also used in transmission systems where cables of various lengths must be precisely equalized [6,22]. In this case the shape of the equalization does not change while the amount of the equalization is varied. The previous generations of equalizers used passive RLC, active-RC and hybrid bipolar technologies using thin and thick-film techniques.

An amplitude equalizer is characterized by the following second order transfer function:

$$H(S) = \frac{V_o}{V_{in}} = \frac{S^2 + B\left[1 + A_1(1-K)\right]S + \omega^2}{S^2 + B\left[1 + A_1 K\right]S + \omega^2} \tag{6}$$

Three design parameters are of interest, namely $\omega_0$, B and K. $\omega_0$ controls the center frequency, B controls the bandwidth (or $Q$) and K controls the bump height (or gain). We have succeeded to develop the fully-programmable MOSFET-C architecture of Fig. 4 where independent programmability of these three parameters is possible [23]. The development of the new structure started with a block diagram representation of equation (5) which is implemented by the proper interconnection of basic linear cells such as those in Fig. 1. The DC control voltages $V_{\omega_0}$, $V_{bw}$ and $V_K$ provide independent tunability for $\omega_0$, B and the bump height respectively. Simulation results using SPICE for $\omega_0$, the bump height and the bandwidth are shown in Figs. 5(a) (b) and (c) respectively.

# 5 CONCLUSIONS

This chapter introduced the new design concept of CAD-compatible analog system design. The concept is based on the interesting connection which we pointed out between programmability, CAD design and physical layout of analog integrated circuits. It achieves design parameterization using programmability resulting in a fixed physical layout for the analog circuit or system. The idea was demonstrated in the design of MOSFET-C filters and amplitude equalizers. The work is an attempt to brigde the gap between circuit design and CAD-oriented efforts in the analog field and bring analog VLSI design a step closer towards automation.

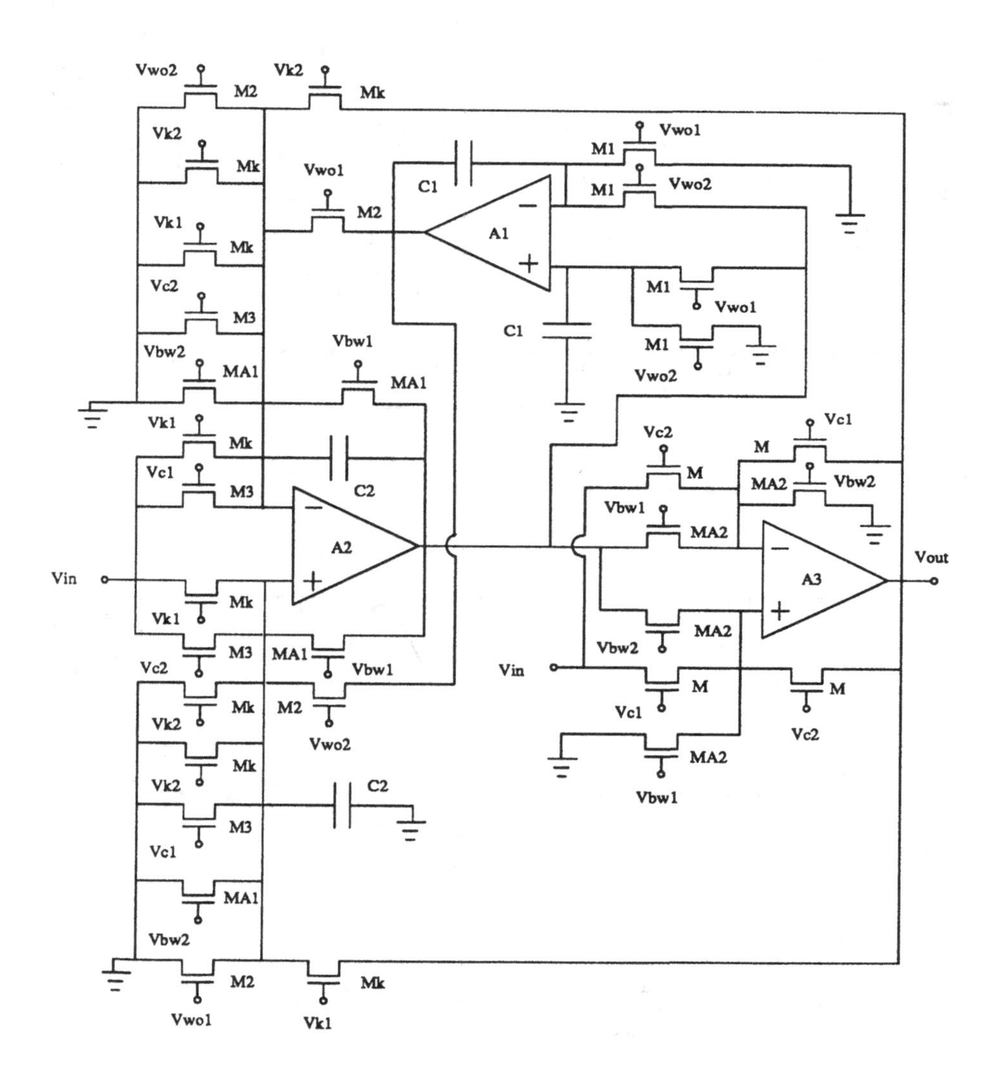


Figure 4: MOSFET-C implementation of a programmable Bump equalizer.

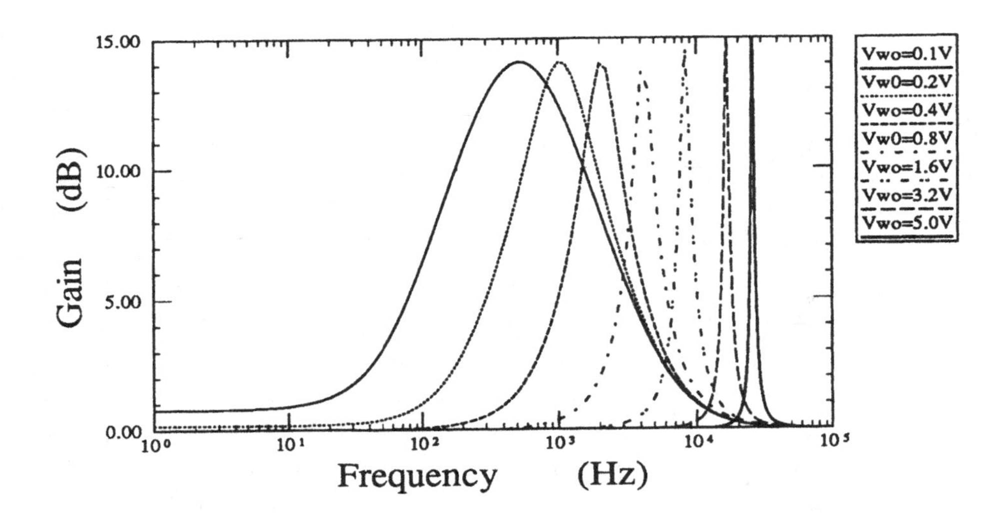


Figure 5(a): $\omega_0$ programmability.

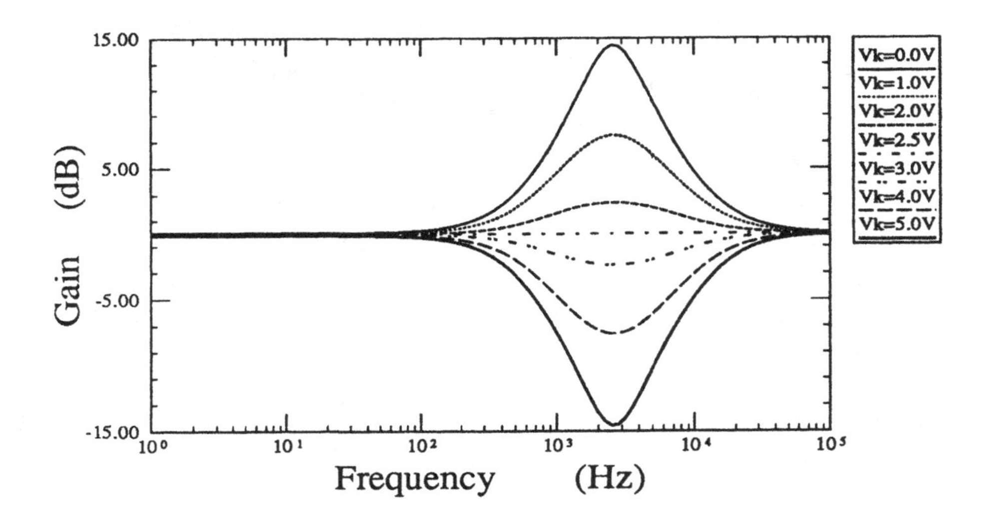


Figure 5(b): Bump height programmability.

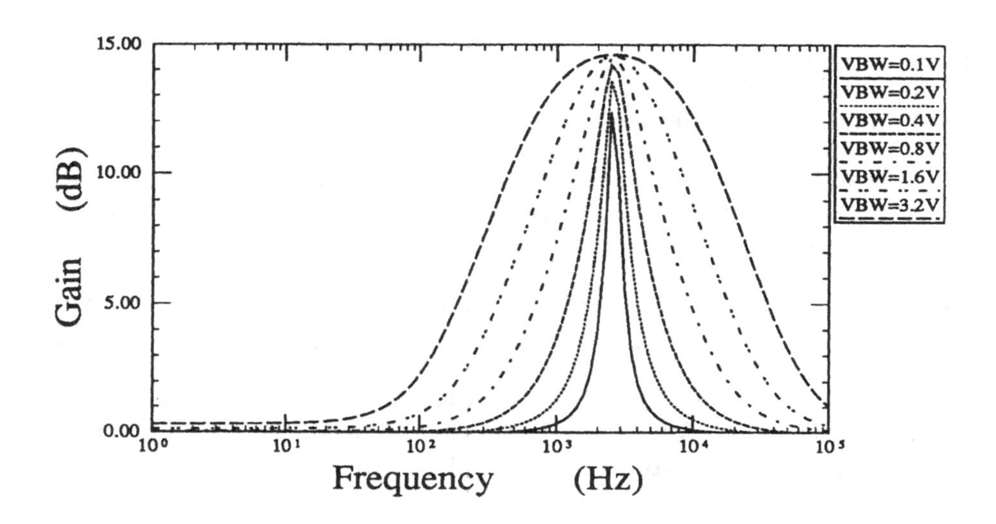


Figure 5(c): Bandwidth programmability.

# References

[1] P. E. Allen, et al., "AIDE2: An Automated Analog IC Design System," Proc. IEEE Custom Integrated Circuits Conference, pp. 498-501, May 1985.

[2] M. Ismail "Reconfigurability, Versatility and Modularity in Analog IC Design," presented at the Semiconductor Research Corporation (SRC) Workshop on Analog Design Automation, December 2, 1988.

[3] J. L. Hilbert, proc. SRC Workshop on Analog Design Automation, Melbourne, FL, December 1988.

[4] H. Y. Koh, C. H. Sequin and P. R. Gray, "Auto Synthesis of Operational Amplifiers Based on Analytic Circuit Models," proc. IEEE ICCAD, pp. 502-505, November 1987.

[5] L. R. Carley and R. A. Rutenbar, "How to Automate Analog IC Designs," IEEE Spectrum, pp. 26-30, August 1988.

[6] J. F. Duque-Carrilo, et al. ,"Programmable Switched-Capacitor Bump Equalizers," to be published.

[7] M. Ismail, S. Smith and R. Beale, "A New MOSFET-C Universal Filter Structure for VLSI," IEEE Journal of Solid State Circuits, Vol. 23, pp. 183-194, February 1988.

[8] Y. Tsividis, et al., "Continuous-time MOSFET-C Filters in VLSI," IEEE Trans. Circuits and Systems, Vol. CAS-33, pp. 125-129, February 1986.

[9] M. Ismail, "Four-Transistor Continuous-time MOS Transconductor," Elec. Letters, Vol. 23, No. 20, pp. 1099-1100, September 1987.

[10] M. Ismail and D. Ganow, "MOSFET-Capacitor Continuous-time Filter Structures for VLSI," Proc. of the IEEE International Symp. on Circuits and Systems, pp. 1196-1200, May 1986.

[11] M. Ismail and D. Rubin, "Improved Circuits for the Realization of MOSFET Capacitor Filters," Proc. of the IEEE International Symp. on Circuits and Systems, pp. 1186-1189, May 1986.

[12] M. Ismail and J. Prigeon, " A Novel Technique for Designing Continuous-time Filters in MOS Technology, " Proc. of the IEEE International Symp. on Circuits and Systems, pp. 1665-1668, June 1988.

[13] M. Ismail, "Continuous-time Analog Design for MOS VLSI," State-of-the-Art Review invited paper, Proc. of the 30th Midwest Symp. on Circuits and Systems, pp. 707-711, August 1987.

[14] S. Dupuie, S. Bibyk and M. Ismail, "A Novel all-MOS High-speed Continuous-time Filter," Proc. of the IEEE International Symp. on Circuits and Systems, pp. 675-680, May 1989.

[15] C. Mead and M. Ismail, *Analog VLSI Implementation of Neural Systems.* Kluwer Academic Publishers, Boston, MA, 1989. Ch. 5.

[16] F. Salam, N. I. Khachab, M. Ismail and Y. Wang, "An Analog MOS Implementation of the Synaptic Weights for Feedback Neural Networks," Proc. of the IEEE International Symp. on Circuits and Systems, pp. 1223-1225, May 1989.

[17] M. Ismail et.al, "Continuous-time Analog Signal Processing Techniques for MOS VLSI," Proc. of the First SRC TECHCON, Dallas, TX, pp. 358-361, October 1988.

[18] N. I. Khachab and M. Ismail, "A MOS Multiplier/Divider Cell for Analogue VLSI," Elec. Letters, Vol. 25, No.23, pp. 1550-1552, November 9, 1989.

[19] N. I. Khachab and M. Ismail, "Novel Continuous-time all MOS Four Quadrant Multipliers," Proc. of the IEEE International Symp. on Circuits and Systems, pp. 762-765, May 1987.

[20] E. Suckhinger and W. Guggenbuhl, "A Versatile Building Block: The CMOS Differential Difference Amplifier," IEEE J. Solid-State Circuits, Vol. 22, pp. 287-296, April 1987.

[21] N. I. Khachab and M. Ismail, "A New Technique for the Implementation of Programmable-Q MOSFET-C Filters," Proc. of the 32nd Midwest Symp. on Circuits and Systems, August 1989.

[22] R. R. Cordell, "A New Family of Active Variable Equalizers," IEEE Trans. Circuits and Systems, Vol. CAS-29, pp. 316-322, May 1982.

[23] S. Sakurai, E. Sanchez-Sinencio and M. Ismail, "Continuous-time MOSFET-C Bump Equalizers," to be published.

[24] S. Zarabadi and M. Ismail, "Continuous-time Analog Cells for High-speed Signal Processing," to be published.

## ACKNOWLEDGEMENTS

The author would like to thank the U.S. National Science Foundation and the U.S. Semiconductor Research Corporation for their support. The help of Mr. Nabil I. Khachab of the Ohio State University in preparing this chapter is greatly appreciated.

# INDEX


Analog, Building blocks, 82
, Layout Generation, 102, 132
, Standard Cell, 140
, Sub-blocks, 56
Analog Circuit, Hierarchy, 9, 54, 80, 158
, Symbolic simulation, 85
Analog Design, Automation, 3, 79, 98, 164
, for CAD, 164
, Design procedures, 79
, Domain presentation, 23
, Engineer task profile, 4
, Environment, 30
, Expert systems, 36
, Methodology, 81, 171
, Module generation, 66
, Optimization, 12, 21, 97, 100
, Paradigm, 17, 18
, Space exploration, 71
, Strategy, 16
, style selection, 11
, Tools, 113
, Translation, 56, 58
Analog Design, Conceptualization, 1
Analog Integrated Circuit Design, 6, 138
Analog VLSI, 163
Analytic Module Generation, 92
Approximation Functions, 153
ASIC, 75, 97, 114
CAD Tools, 46, 80, 151
CAD-Compatible Analog Design, 163, 164
Cell, Characterization, 119
, Design, 124
Characterization, 115, 119
Computer-Aided Techniques, 46

Continuous-time, Analog MOS Integrated Circuits, 165
, Analog MOS Systems, 163
, Filters, 159
Design Tenets, 7
Expert System, 29
Fixed Analog Cells, 165, 168
Fixed Cells, 163
Global Optimization, 21
ISDN, 138
Knowledge-based Expert System, 33
Ladder Design, 158
Layout Editor, 137
Layout Synthesis, 159
MOS Resistors, 165
MOSFET-C Filters, 171, 175
Netlist, Evaluation, 159
, Synthesis, 157
Numerical Simulation, 86
Operational Amplifiers, Automated Design, 45
, CMOS, 67, 75, 94
Parameterized Analog Cells, 29, 163
Programmable Cells, 163
Rule Based Design, 143
Silicon Compiler, 151
Simulated Annealing, 100
Switched-Capacitor Filters, 151
Symbolic Analysis, 85
Symbolic Simulator, 87